MUSEUMS AND THE CLIMATE CRISIS

Museums and the Climate Crisis shows how museums can respond to the interrelated global climate, biodiversity and pollution crises. They have a unique role because they take a long-term perspective, and their scholarship and independence mean that thesy remain trusted by the public.

Providing insights and international case studies from a range of museum and gallery professionals, academics and consultants, this book explores how museums can use this unique perspective to engage the public as active citizens, and how they are exemplars of good practice in areas such as emissions reduction and encouraging biodiversity. It shows how museums can combat climate exhaustion by drawing on understandings about positive motivation, and how to develop exhibitions, events and activities that motivate visitors to take action. Taking a broad approach beyond purely climate issues, the contributions touch on the use of renewables, environmental controls and standards, travel (including virtual couriering), waste management (including recycling, plastic reduction and composting), reducing pollution and increasing biodiversity within museums.

Museums and the Climate Crisis will be important reading to those studying in the fields of Museum Studies, Heritage Studies and Conservation. Taking a practical approach, it will also be beneficial to museum, gallery and heritage professionals who are grappling with the challenges of the climate crisis.

Nick Merriman is Chief Executive of the Horniman Museum and Gardens in London and an Honorary Professor at University College London and the University of Manchester.

MUSEUMS AND THE CLIMATE CRISIS

Edited by Nick Merriman

Routledge
Taylor & Francis Group
LONDON AND NEW YORK

Designed cover image: Jamie Craggs for the Horniman Museum and Gardens

First published 2024
by Routledge
4 Park Square, Milton Park, Abingdon, Oxon OX14 4RN

and by Routledge
605 Third Avenue, New York, NY 10158

Routledge is an imprint of the Taylor & Francis Group, an informa business

British Library Cataloguing-in-Publication Data
A catalogue record for this book is available from the British Library

ISBN: 978-1-032-38943-1 (hbk)
ISBN: 978-1-032-38941-7 (pbk)
ISBN: 978-1-003-34760-6 (ebk)

DOI: 10.4324/9781003347606

Typeset in Times New Roman
by Apex CoVantage, LLC

For Maria

CONTENTS

FIGURES

CONTRIBUTORS

Zehra Ahmed is the Sustainability, Accessibility and Diversity and Inclusion Projects Coordinator at the Australian Museum in Sydney, Australia.

Maria Balshaw has been Director of Tate since June 2017. Previously, she was Director of the Whitworth, University of Manchester; Director of Manchester City Galleries; and Director of Culture for Manchester City Council.

Jo Beggs is currently Head of Development at Manchester Art Gallery. Her career has spanned museums and galleries, theatre, festivals and hospitality.

Lisa Broadest is Head of Operations at Leeds Museums and Galleries and the organisation's Green Champion.

Victoria Burns has 25 years' experience of working in arts and heritage as an organisational development consultant and creative producer. She is Co-Director of Climate Museum UK and National Coordinator for Culture Declares Emergency.

Eduardo Carvalho is Brazilian Curator and Cultural Manager working in global projects that use culture, technology and design to engage the audience about science and climate emergency

Simbarashe Shadreck Chitima is Lecturer in the Department of Archaeology, Cultural Heritage and Museum Studies at Midlands State University, Zimbabwe, with research interests in museum greening, conservation, accessibility, education and exhibition design.

Carole Destre is the Climate and Biodiversity Coordinator at the Horniman Museum and Gardens in London.

John Fraser is Conservation Psychologist, Architect and Educator. He is President and CEO of Knology and also Editor of *Curator: The Museum Journal*.

Yvonne Hardman is the Head of Collections and Programmes at Leeds Museums and Galleries.

Clare Matterson CBE was Executive Director for Engagement at the Natural History Museum in London from 2018 to 2022. She is currently Director General at the Royal Horticultural Society, London, the UK.

Bridget McKenzie is Creative Curator and Consultant dedicated to Regenerative Culture. She has been Director of Flow Associates since 2006 and in 2018 established Climate Museum UK and co-founded Culture Declares Emergency.

Nick Merriman is Chief Executive of the Horniman Museum and Gardens in London and an Honorary Professor at University College London and at the University of Manchester, Manchester, the UK.

Jenny Newell is Curator for Climate Change, Climate Solutions Centre at the Australian Museum in Sydney, Australia.

Richard Piacentini is the President and CEO of Phipps Conservatory and Botanical Gardens in Pittsburgh, the United States.

Kathryn Simpson is Policy and Projects Manager at the UK National Museum Directors' Council. She has worked in policy, strategy and international relations for over 13 years in both the cultural sector and central government.

Caitlin Southwick is Founder and Executive Director of Ki Culture and Sustainability in Conservation (SiC), Amsterdam, the Netherlands. Before this, she worked in the conservation field for eight years.

Sarah Sutton is CEO of Environment and Culture Partners (ECP), a United States-based non-profit accelerating national and global environmental leadership in the culture sector.

Alison Tickell established Julie's Bicycle in 2007 to help the UK music industry reduce its environmental impacts and has since extended its remit to the wider creative and cultural communities.

Dean Whiteside is Head of Operations and Buildings at the Whitworth and Manchester Museum in the University of Manchester.

Douglas Worts is Culture and Sustainability Specialist, living in Toronto, who has worked in the museum world for 45 years, including 25 years at the Art Gallery of Ontario, Toronto, Canada.

INTRODUCTION

With the situation regarding the climate and biodiversity crisis changing almost on a weekly basis, it could be argued that a book on the topic isn't the right medium for dealing with such a crucial subject. An edited volume takes many months to compile and publish, by which time – it might be said – the information is already out of date. However, while it is true that the picture is a changing one, it is also the case that most of us are working on long-term plans to tackle the crisis, such as reaching net zero carbon emissions by 2030, 2040 or 2050. Approaches to the crisis thus operate at different timescales: there is the need to take urgent mitigating actions, while at the same time understanding that the effect of these actions will play out over the long term.

The contributions to this book are aimed at this longer timescale – one which museums are well-placed to operate within, as institutions that take a long-term perspective. While the details of some of the statistics presented in its chapters may soon be dated, the generalities of what they say, it is hoped, will remain relevant for much longer and be of use to museum and heritage professionals and students who are trying to understand how best to tackle the existential crisis we are all facing.

This book has arisen from discussions that have taken place over the last few years in a variety of fora, including the UK's National Museum Directors' Conference (NMDC) Environment and Ecology Working Group, which has held regular video conferences to share information and practice about museums and the climate crisis. A particular spur was the in-person NMDC conference 'Museums and Galleries Responding to the Climate and Ecological Crisis' conference held on 7–8 March 2022 at the Whitworth in the University of Manchester. A number of the contributors to that conference have kindly revised and updated their presentations for this book.

In soliciting contributions for this book, I drew on my own experience, and for this reason many of the papers are by UK authors. I was also guided by advice from a number of colleagues when casting the net more widely internationally, seeking in particular to have some contributions from outside the UK and North America. Unfortunately, one major contribution from South Asia was not forthcoming due to illness, which means the book is slightly less international in coverage than I had ideally wished.

Structure of the book

The book is divided into two principal sections. Part 1, entitled 'The Big Picture', consists of ten chapters which look in depth at particular issues. These range from broad overviews of the whole field of museums and galleries and the climate and biodiversity crisis (Balshaw, Merriman, Simpson), how best to encourage the public to take action in relation to the climate crisis (McKenzie and Burns, Sutton and Fraser), challenges in conservation and collections care (Southwick), a conceptual framework for how museums can change the wider culture (Worts) and overviews of experience in particular countries or organisations (Carvalho, Chitima, Tickell).

Part 2 consists of case studies from particular museums or related institutions that exemplify interesting and leading practice. Chapters are provided on two of the world's leading museums with natural history collections (Ahmed and Newell, Matterson), a pioneering art gallery (Beggs and Whiteside), an innovative conservatory and botanical garden (Piacentini), an early-adopting regional museum (Broadest and Hardman), and a prize-winning London museum and gardens (Destre and Merriman).

ACKNOWLEDGEMENTS

I would like to thank all of the contributors to this book for agreeing in the first place to write something, for their hard work in preparing their papers and for their patience while all of the chapters have been assembled and edited.

In terms of individuals, I would like to thank Beth Reynard (formerly of NMDC), Kathryn Simpson and Suzie Tucker (of NMDC) for all of their support in this work. Colleagues at the Horniman Museum and Gardens have been an inspiration in all that they do in this area and have supported me to finally deliver this book. I would also like to thank Heidi Lowther and Manas Roy at Routledge for their advice, feedback and patience.

PART 1

The big picture

1

THE ROLE OF MUSEUMS AND GALLERIES IN ADDRESSING THE CLIMATE AND ECOLOGICAL CRISIS

Nick Merriman

The separation of nature and culture in museums

In cabinets of curiosities, natural and cultural collections were intermingled in a holistic view of the world (Impey and MacGregor 1985). With the Enlightenment, Judeo-Christian thought separated nature and culture, allowing nature to be exploited, dominated and controlled – to be seen as a hostile encumbrance to progress and a resource to bring riches (Watson 2012). This division of nature from culture has had a fundamental impact on the development of museums by leading to the separation of natural history from cultural collections into their own dedicated galleries or museums. For example, showing its origin in Hans Sloane's cabinet, the British Museum's collection combined natural and human history from its foundation in 1753 until the former was separated to establish the Natural History Museum in 1881, and, within large regional organisations like Manchester Museum, natural history has always been presented in separate galleries, often some distance from displays of human cultures which lived in the same environments.

For most of the time since its emergence as a distinctive critical field in the 1980s, museum studies has primarily been a human-centred subject, emerging as it did from the disciplines of history, archaeology, anthropology and art history, with their focus on human activity, material culture and critical theory. Those who worked on natural sciences in museums were working within a scientific paradigm that had little in common with museological theory, but were often doing vital practical work to highlight species decline, campaigning about the Red List of endangered species and gathering evidence for change and loss, using the incredible databank of biological diversity in natural history collections. Similarly, environmental education was, and is, seen as a vital function of museums, but for much

DOI: 10.4324/9781003347606-2

of its history it employed a top-down deficit model, which thought that education would lead to appreciation, which in turn would lead to care and hence preservation of species and their habitats (see, e.g. Oliver 1971).

The natural environment and natural history were of course the subject of some early critical studies, such as Donna Haraway's classic article on 'Teddy Bear Patriarchy' (1984), but they tended to critique representations of the natural world from the perspective of what they said about the society that had produced them. Georges Henri Rivière and Hugues de Varine's concept of the ecomuseum, articulated in 1971 and subsequently developed (de Varine 1985), was to a large extent about human activities taking place within landscapes – mostly industrial ones – with an emphasis on preservation, often with a strongly nostalgic component (Davis 1999). Again, the main focus was on people, but more recently, the ecomuseum movement has begun to lead the response by museums to the climate and ecological crisis (Borrelli et al. 2022).

The idea that nature and culture can be seen as separate realms has however long been criticised, partly because we cannot find anywhere on Earth, including the deep oceans, that has not been affected by people, particularly in terms of plastic and particulate pollution. Some scholars have suggested that 'nature' should be retitled as the infelicitous 'socio-environmental arrangements' (Swyngedouw 2013) or, perhaps better, as 'seeming nature' (Gandy 2002: 110).

The historic nature–culture divide has affected all areas of museology, including the physical structures of museums, and different paradigms of thinking between humanities and scientific disciplines in museums and has led to a blind spot in museological thinking that is only recently being addressed.

The divide has led, first, to a partial account of human history in museums. Accounts of slavery, colonial exploitation and the Industrial Revolution downplay the devastation of environments through deforestation, crop planting, mining and the slaughter of animals for the leather, horn, ivory and feather products we see in our museums (Das and Ghadiali 2021). Products made or consumed in the UK from materials from abroad (sugar, tea, coffee, rubber, oils, furniture, jewellery) are separated from natural history galleries which present them as pristine products, separate from any supply chain, and they are also presented separately from anthropology galleries showing the peoples from those regions.

The rise of Europe as an industrial power in the 18th and 19th centuries through to its height before the Second World War was based on the taming and exploitation of the natural resources of its own lands and especially of foreign territories. The great majority of the goods that people bought, many of them now shown in our local history museums, were made from natural materials from overseas. But because of this separation between seeming nature and culture, our museums show the products, innocently sitting on showcase shelves, as examples of industrial production and consumption, and they might sometimes show the animals and plants from which they were derived, but they are usually silent on the processes that led from one to the other.

This silence has led to the other, much greater, consequence of the separation between nature and culture in museums: our slowness in confronting the enormity of the environmental and ecological challenges facing the planet at the moment. Challenging the separation between nature and culture in museums also forces the realization that environmental and social justice are inextricably bound up with one another. Put plainly, the Global South emits far lower carbon emissions than the North, yet suffers disproportionately from the effects of the North's historic emissions, and in the Global North, it is the disadvantaged citizens – the poor, ethnic minorities, those with health issues – who suffer most from the consequences of the environmental crisis (Thunberg 2022). Museum literature has focused intently on social justice issues, pointing to the exclusionary culture and practices of museums and highlighting ways of removing barriers, but very little has explored linkages with climate justice issues. Because of the sector's blind spot in relation to the climate and biodiversity crisis, this book focuses primarily on rectifying this; however, the interrelationship with social justice must permeate any account.

As a reminder of what we are facing, Table 1.1 summarises some of the latest scientific evidence.

So, the climate and ecological emergencies and the pollution of air and sea add up to an existential threat to us all, including that to museums and the collections which distinguish them, to their buildings, to the people who work in them and to their visitors.

But in museums, we have said and done relatively little about this until recently. The exception has been Robert Janes, the leading authority in relation to museums and the climate emergency. Some 15 years ago, his book *Museums in a Troubled World* (Janes 2009) provided a passionate and prescient plea for museums to engage with issues such as climate change and environmental degradation to avoid becoming totally irrelevant. His latest book *Museums and Societal Collapse* provides a stark warning of the consequences of inaction (Janes forthcoming). Other than this, most thinking and action have come either from within academia (e.g. Harrison and Sterling 2021) or from groups outside the museum mainstream. Of the latter, along with wider groups such as Extinction Rebellion and more specific groups such as Culture Declares Emergency, activists and protesters have been raising the profile of climate and ecology and forcing museums to look more squarely at it. For example the New-York-based Climate Museum and the Climate Museum UK are online and pop-up museums creatively stirring responses to the climate and ecological emergency (Climate Museum n.d.; Climate Museum UK n.d.).

More recently, a range of books and articles on museums and climate change have appeared with their focus on different parts of the world (Garthe 2023; Neilson and Cameron 2016; Newell et al. 2017; Sloggett and Scott 2023; Sutton 2022), and an issue of *Museum Management and Curatorship* (Davis 2020) has been dedicated to the climate crisis. Henry McGhie has valuably linked museums to the achievement of the UN's Sustainable Development Goals (McGhie 2019).

TABLE 1.1 A summary of global climate and biodiversity challenges

- Scientists agree that we are in the middle of the sixth mass extinction phase in the planet's history, where the rates of extinctions are far outpacing the background rate. The last mass extinction was in the Cretaceous, 66 million years ago (Kolbert 2014).
- The current geological phase has been named the Anthropocene due to the visible impact of humans on the geological record.
- The last eight years were the eight hottest years on record around the world (NASA 2022).
- The UN Intergovernmental Panel on Climate Change (IPCC) has warned that global warming of 2°C above pre-industrial levels could lead to increased droughts and flooding, sea level rises, ecosystem change, species loss and extinction on land and in the sea, reduced productivity for agriculture and fishing and climate-related poverty and disease (IPCC 2023).
- Climate change is likely to be the greatest cause of species extinctions this century. The IPCC reports that increases in global temperatures threaten large-scale extinctions of species and whole ecosystems (IPCC 2022).
- Antarctica is losing ice mass (melting) at an average rate of about 150 billion tons per year, and Greenland is losing about 270 billion tons per year, adding to sea-level rise (NASA 2023).
- Coral reefs harbour the highest biodiversity of any ecosystem globally and directly support over 500 million people worldwide, mostly in poor countries. But they are among the most threatened ecosystems on Earth, largely due to unprecedented global warming and climate changes, combined with growing local pressures. We have lost over half of all coral reefs since 1950 (Eddy et al. 2021).
- Between 12% and 20% of global greenhouse emissions come from tropical deforestation – burning forests releases huge amounts of CO_2 to the atmosphere, contributing to climate change (Watson and Schalatek 2020).
- Five hundred million people live in areas that experience desertification – this degraded land is increasingly affected by climate change (IPCC 2019).
- People numbering 2.3 billion already live in countries experiencing high water stress – climate change will make their lives harder (United Nations 2021).
- In 2019, 99% of the world's population was living in places where the World Health Organization (WHO) air quality guidelines levels were not met. The combined effects of ambient air pollution and household air pollution are associated with 6.7 million premature deaths annually (WHO 2022).
- Eighty-four per cent of young people aged 16–25 years across ten countries are worried by climate change (Hickman et al. 2021).
- We are the first generation to know we are destroying the world, but we could be the last that can do anything about it (WWF 2019).

Despite this, museums themselves have mostly been slow to respond. Given the enormity of the crisis, it would be easy to say that their slowness is a sign that they have little to offer in mitigation. However, museums, in fact, have a unique role to play in facing the climate and ecological crisis.

Along with archives, and other heritage organisations, museums are institutions of the long term, mandated to think beyond short-term political and economic cycles. Many museums hold palaeontological collections documenting the five previous mass extinctions and are only too aware that we are living in the sixth one, the Anthropocene. A major part of a museum's role is to preserve its collections for posterity and in perpetuity. For the entire history of museums, the assumption has been that the future – while being technologically different – would be mostly similar in terms of environment. Now, however, we cannot guarantee that environment, society and museums themselves will exist in the ways we have taken for granted.

Museums are also among some of the few institutions that retain the public's trust. The annual IPSOS Veracity Index (2022) regularly shows museum curators in the list of top six most trusted professionals alongside nurses and teachers, and trends in the United States reveal a similar level of public confidence (Merritt 2019). They are also very popular: in England, in 2019–2020, 51% of people aged 16 and over had visited a museum or gallery at least once in the past year (DCMS 2020). Museums might reasonably be described as a kind of slow form mass media.

As a consequence of their long-term perspective, the existential threat they are under, their public trust and their popularity, museums have a moral and ethical responsibility to highlight the climate and ecological crisis and engage their audiences in positive actions to mitigate it. Through their collections and their spaces, they are also particularly well placed to bridge the division between climate change and biodiversity issues, which has existed in the public sphere ever since the UN Climate Change Convention and the UN Convention on Biological Diversity were both established in 1992, with their own separate Conferences of the Parties (COPs).

What are museums doing?

Museums have focused on two broad areas of activity, characterised by Douglas Worts in this volume as 'inside' and 'outside'. Inside actions are about getting one's own house in order and being an exemplar of good practice, while outside actions are those that engage the public in becoming active citizens in fighting the crisis. Museum organisations have all now, albeit belatedly, recognised the urgency of the climate and biodiversity emergency. The UK Museums Association has developed a campaign and resources on Museums for Climate Justice (Museums Association n.d.) and the National Museum Directors' Council has an Environment and Ecology Working Group, which shares best practice and organises conferences, as do the American Association of Museums' Environment and Climate Network and the ICOM Sustainability Working Group. Starting in the commercial art sector, the Gallery Climate Coalition now has over 800 members worldwide focusing on practical ways to lower the impact of the global art world. In the wider cultural sector, Julie's Bicycle has been working for two decades to provide support and advice (Tickell, this volume). Indeed, it can sometimes seem that there is an overwhelming amount of advice out there, and, as a result, finding one's way among the mass of information can be daunting.

Getting the house in order: issues and challenges

This volume shows examples of how museums have been pursuing their ambitions to reduce their greenhouse gas emissions, increase biodiversity and reduce waste and pollution. Most have declared that they will achieve 'net zero' emissions by a certain date. Many of the initiatives are common to all organisations, irrespective of their sector, such as improving thermal efficiency and transitioning away from natural gas. Sector-specific guidance such as the Arts Green Book on sustainable buildings (Buro Happold and Renew Culture 2022) has been made freely available, and professional organisations are now providing clear advice on actions and monitoring.

Many museums are also tackling issues that are mainly the preserve of the sector itself, such as the parameters for environmental control needed to ensure the preservation of the collections and resource-hungry issues such as air freight for international loans. As Southwick (this volume) points out, we now have an unsustainable situation whereby standards developed originally for the care of paintings in London in the 1960s have become blanket environmental conditions for all collections worldwide, resulting in the installation of air conditioning to meet them for loans, when lending institutions do not meet them themselves. A large part of the problem here is the one-way rachet of professional standards, which rewards the tightening of controls. Leadership is needed at all levels to tackle this, as noted by Balshaw (this volume), and solutions need to be suitable for each specific context. Not all collections need the same conditions, and different parts of the world have their own ways of managing environments, based on traditional architecture and practices (e.g. Chitima, this volume).

Pollution, waste and biodiversity are fundamental issues in the global crisis, yet they tend to be overshadowed by climate. Museums, with their collections, visitors and green spaces, can do something to rebalance, and there are good examples in this volume of initiatives around 'greening', tackling air pollution and decreasing plastic waste, with many of the good practice initiatives also being an opportunity for active public engagement. Leeds Museums & Galleries have been pioneers in the development of sustainable exhibition-making, including recycling of materials used to produce displays (Broadest and Hardman, this volume). The Design Museum's exemplary exhibition 'The Waste Age' was accompanied by a detailed analysis of its carbon footprint which showed, among other things, that 50% of emissions were caused by shipping one artwork from Ghana, and the largest single carbon footprint in the exhibition build comprised 4,800 steel screws which had an impact of around 7% of the total exhibition footprint (Design Museum and Urge Collective 2023).

Although the museum sector is a very small one in terms of carbon emissions, for the reasons set out earlier, its voice is an influential and important one. So, practicing what one preaches becomes vital to credibility and authenticity. A series of key issues and challenges has arisen from this. The first is about growth in the museum sector. The often unspoken assumption of museum boards and staff, funding bodies, politicians and other stakeholders in the decades since the Second World War has been that a healthy museum is one that is constantly growing its collections, its visitor numbers, its digital interactions, its buildings, its staff and its budgets. All the metrics by which museums are measured are quantitative ones, with the expectation that business plans will project annual increases in everything measured. Professional reputation is based on size and growth, measured in new buildings, acquisitions or increased funds. No one is ever rewarded in the museum sector for making the collection or estate smaller, even if this might be more sustainable.

COVID-19 provided a challenge to this model of constant growth, with buildings closed and visitor numbers vastly reduced. This could have provided an opportunity to reset expectations about sustainability, with museums embracing new forms of operation, whether it be about 'de-growth' (Latouche 2009)) or sustainable alternatives such as that exemplified by the economist Kate Raworth in her 'Doughnut Economics' (Raworth 2017). Certainly, if museums are to tackle the challenges of the climate and biodiversity crisis, they will have to re-examine their expectations about constantly growing and construct new measures of success, for example around quality of visitor experience, diversity of audiences and sustainability of operation.

A second issue relates to some confusion about what 'net zero' means in both practice and impact. It is often assumed that if we reach 'net zero' by a certain date, global warming will remain less than 1.5°C above pre-industrial levels, and the job will have been achieved. However, what net zero actually means is that – in the year it is achieved – the volume of CO_2e emissions in the atmosphere will be balanced by the volume of emissions which are sequestered through living organisms or methods such as carbon capture and storage. It doesn't tackle the previous two centuries of emissions which are still having an impact on the environment (Fankhauser et al. 2022). So, even when net zero is achieved, this existing impact has to be mitigated: the ice sheets will not stop melting when we reach net zero.

Most future museum net zero scenarios include some residual emissions from sectors that will find it difficult or impossible to reach gross net zero (such as production of plastics), and, as a result, they anticipate reaching net zero through some form of off-setting or carbon-positive measures. Many of these have been criticised for being unethical, or at the very least inappropriate, given the length of time it will take for trees to grow to maturity (Krishnan et al. 2023). The Gallery Climate Coalition has suggested using funds that would have been given to offsetting schemes for investment into 'Strategic Climate Funds' to support initiatives to reduce the art world's carbon footprint or to support projects, making a difference to the climate crisis (GCC 2023).

The other ethical issue that continues to bedevil museum climate initiatives is that of fossil fuel sponsorship. Many prominent museums and galleries have cancelled their association with oil, gas and coal companies because of a combination of ethical concerns about accepting such income and because of mounting activist protests. Others argue that such companies are investing in renewables as a matter of business adaptation and that they have to be part of future solutions. The IPCC report however points out that 'Public and private finance flows for fossil fuels are still greater than those for climate adaptation and mitigation' (IPCC 2023: 11). As a result, many feel that assertions that fossil fuel companies are an important part of the solution ring hollow in the face of a growing crisis (Lawson 2022).

Alongside these ethical concerns about getting the house in order are huge practical ones. With the advice abounding in the sector, and with many developing roadmaps to get them to net zero, most museums know what measures they should be taking to get there. However, insulating old buildings and installing heat pumps and photovoltaics come at great cost, and at present there are no reliable sources of funding for this work. In addition, many cultural buildings are protected by historic building legislation, which currently means that measures such as double-glazing and solar panels are deemed unacceptable.

The short-term thinking of politics and economics means that it always seems easier to kick the can down the road in relation to these huge issues: climate always seems to be someone else's problem. In the UK, the Theatres Trust has estimated that £1.1 billion is needed to make theatres sustainable and accessible and that lack of finance is by far the largest obstacle to doing so (Jowett 2021). Until governments provide funding and review planning controls, museums will be unable to meet their decarbonisation targets.

Engaging the public as active citizens

While having an important role as exemplars of good practice in terms of decarbonisation and tackling pollution, waste and biodiversity, the real long-term impact of museums is likely to be through their role in engaging a large and wide public which trusts what they say. The big challenge is how to do this in a way that spurs audiences to action, rather than exacerbates their feelings of powerlessness and leads to inaction.

Most museums have tended to operate in traditional broadcast mode in relation to the climate crisis, alerting people to the facts in the hope that the gravity of the situation will galvanise people. This has been termed the 'deficit model' of science communication (Miller 1983) and has long been criticised for not understanding the complexities of the context of communicating in public places (e.g. Gregory and Miller 1998).

There is now a growing literature in the field of environmental psychology, which highlights something of the complexity of the issues (Steg and de Groot 2018). Spence and Pidgeon (2009) provide a useful early overview of approaches to climate communication. They note, as above, that there is an assumption in approaches to communicating about the climate crisis that when people are provided with information, they will change their behaviour in a positive way.

However, they cite research on the 'contemporary discourse of fear' surrounding climate change, which can have the effect of turning people off.

> The psychological literature on this point indicates substantial evidence, from domains such as health protection, that fear framing will initiate action as long as individuals feel that they have some degree of control to act in response to the problem.
>
> *(ibid.: 9)*

They recommend that 'climate communicators should therefore seek to frame emotive messages alongside positive, credible steps which people themselves can take' (*ibid.*). *The Handbook of Climate Change Communication* (Filho et al. 2018) and *the Oxford Encyclopaedia of Climate Change Communication* (Nisbet et al. 2018) provide useful guidance on ways of framing successful communication, and McGhie et al. (2018) are one of the museum examples.

Building on this context, Sutton and Fraser (this volume) provide a good overview of both the literature around climate change communication, with some examples of current practice in the United States. This includes the remarkable example of the National Network for Ocean and Climate Change Interpretation (NNOCCI), which has grown into a Community of Practice of over 40,000 trainees and develops new techniques, applications and content. The Australian Museum has probably the most comprehensive approach to climate communication, as described by Ahmed and Newell (this volume), underpinned by large-scale audience research and segmentation, which enable the museum to create a range of approaches to different audiences, using exhibitions, videos, digital resources and discussions.

Many of the other contributors to this volume highlight innovative ways of engaging people in the climate and biodiversity crisis, particularly involving art and the emotions. As Simpson (this volume) notes, art galleries have been in the forefront of climate engagement, such as Tate's showing of Olafur Eliasson's melting glacier chunks called *Ice Watch* and his subsequent exhibition *In Real Life*. Active involvement through storytelling is vital to motivating the public to action, as described by Carvalho (this volume) and McKenzie and Burns (this volume).

But it's not only artists and stories that engage the emotions: 'scientific' natural history museums also have staged dramatic interventions in their galleries to highlight vital issues. In 2019, curators at Bristol Museums shrouded 32 animals in their permanent display which were extinct or endangered in an intervention called 'Extinction Voices' (Charr 2019), and in the same year the Horniman Museum and Gardens removed most of the fish and amphibians in its aquarium, replacing them with plastic rubbish in an initiative entitled 'Beat Plastic Pollution' (Destre and Merriman, this volume). Matterson (this volume) describes how the Natural History Museum in London has re-thought its corporate mission, which is now to encourage visitors to become 'Advocates for The Planet'.

Engaging people to become active citizens in tackling the climate and biodiversity crisis inevitably leads to concerns about eroding the long-standing feeling that museums should be 'neutral' spaces, and about the relationship between being active and being activist. These concerns have led many museums to be reluctant to address the crisis head-on. One of the main barriers has been what Robert Janes has called museums' 'fallacy of authoritative neutrality', as he says 'lest they fall prey to bias, trendiness and special interest groups' (Janes 2009: 146). This is an important issue for museum trustees and staff: the high public trust in museums shown in surveys derives from the fact that they are long-term institutions with the public good in mind, above the shorter-term interests of politics (but at the same time reliant on these for their funding decisions). As far as boards are concerned, museums are supposed to be non-partisan, unpolitical, even-handed, unbiased and neutral.

However, all critical analyses of museums, historically and today, show that they – like history in general – are not neutral: they have always reflected a partial view, usually in favour of the status quo and the more privileged social classes. Their founding history is political, from General Pitt Rivers (1891), who wanted to teach people that social evolution was the way of the world so that they would not listen to 'scatter-brained revolutionary suggestions' to the proponents of the UK's 1845 Museums Act, who were concerned to alleviate drunkenness by giving people an alternative to the tavern (Lewis 1993). Even today, for example, we are not telling an even-handed story about colonialism and slavery, as the Black Lives Matter movement has reminded us.

As Sutton and Fraser (this volume) note, many museums 'remain mired in three swamps of inaction: the false assumption of competitiveness among museums, the false assumption of neutrality, and the fear of losing trust'.

But why should shouting louder about the climate and ecological crisis be a problem? The science is overwhelmingly clear, and it affects literally everyone: museums, their staff, their visitors. Given such an existential threat, as institutions of the long term, able to place what's going on into a wider context, it is an ethical imperative for museums to shout louder and take action.

Conclusion: an ethical museum ecology

Sleepwalking as we are into the sixth mass extinction and an overwhelming series of climate changes, museums must engage as a sector with these long-term global issues. And because the climate and ecological crisis is so interlinked with exploitation and consumption by humans, and the impacts of these changes are felt disproportionally by the disadvantaged, it is inevitable that museums have to address social justice issues as well. To do this, we need to develop a new ethical museum ecology, based on the following principles:

- As institutions of the long term, museums have a leadership role and an ethical imperative in engaging with the urgency of climate, pollution and biodiversity issues.

- Social and environmental justice issues are interlinked and should be viewed holistically.
- 'Authoritative neutrality' is an illusion and has prevented museums from engaging with critical issues.
- Collectively, museums and their audiences can make a significant impact in addressing these issues through their actions as organisations and individuals and by empowering people as citizens.
- We need a new vision for museums that is carbon and pollution neutral, more biodiverse and based on reward metrics other than growth alone.
- Being carbon and pollution neutral and encouraging biodiversity means doing things radically differently.
- It is better to start the journey and be accused of being hypocritical and imperfect rather than doing nothing and waiting to be perfect.
- The museum of the future – while retaining its collection at its heart – will be more focused on issues than objects.

References

Borrelli, N., Davis, P. and Dal Santo, R. 2022. *Ecomuseums and Climate Change*. Milan: Ledizioni Publishing. https://www.ledipublishing.com/book/9788855268387/ecomuseums-and-climate-change/

Buro Happold and Renew Culture. 2022. *The Arts Green Book: Sustainable Buildings*. London: Buro Happold and Renew Culture.

Charr, M. 2019. Bristol Museum Highlights Extinction Crisis. *Museum Next*, 17 August 2019. https://www.museumnext.com/article/bristol-museum-highlights-extinction-crisis/

Climate Museum. n.d. https://climatemuseum.org/

Climate Museum UK. n.d. https://climatemuseumuk.org/

Das, S. and Ghadiali, A. 2021. Genealogies of the Emergency: A Conversation on Race, Climate and Museums with Subhadra Das & Ashish Ghadiali Facilitated by Rodney Harrison & Henry McGhie. In Harrison, R. and Sterling, C. (eds) *Reimagining Museums for Climate Action*. London: Museums for Climate Action, 60–69.

Davis, J. (ed) 2020. Special Issue: Museums and Climate Action. *Museum Management and Curatorship*, 35(6): 584–708.

Davis, P. 1999. *Ecomuseums: A Sense of Place*. Leicester: Leicester University Press.

DCMS. 2020. *Museums – Taking Part Survey 2019–20*. https://www.gov.uk/government/statistics/taking-part-201920-museums/museums-taking-part-survey-201920

de Varine, H. 1985. The Word Ecomuseum and Beyond. *Museum International*, 37(148): 1–2.

Design Museum and Urge Collective. 2023. *Exhibition Design for Our Time: A Guide to Reducing the Environmental Impact of Exhibitions: Draft for Consultation – March 2023*. London: Design Museum. https://designmuseum.org/exhibitions/waste-age-what-can-design-do/take-action

Eddy, T.D., Lam, V.W.Y., Reygondeau, G., Cisneros-Montemayor, A.M., Greer, K., Palomares, M.L.D., Bruno, J.F., Ota, Y. and Cheung, W.W.L. 2021. Global Decline in Capacity of Coral Reefs to Provide Ecosystem Services. *One Earth*, 4(9): 1278–1285. https://doi.org/10.1016/j.oneear.2021.08.016

Fankhauser, S., Smith, S., Allen, M., Axelsson, A., Hale, T., Hepburn, C., Kendall, J.M., Radhika, K., Lezaun, J., Mitchell-Larson, E., Obersteiner, M., Rajamani, L., Rickaby,

R., Seddon, N. and Wetzer, T. 2022. The Meaning of Net Zero and How to Get It Right. *Nature Climate Change*, 12: 15–21.
Filho, W.L., Manolas, E., Azul, A.M., Azeiteiro, U.M. and McGhie, H. (eds). 2018. *Handbook of Climate Change Communication*. New York: Springer.
Gallery Climate Coalition (GCC). 2023. *Strategic Climate Funds*. https://galleryclimatecoalition.org/usr/library/documents/main/gcc_scf-policy-master-doc_mar23_final.pdf
Gandy, M. 2002. *Concrete and Clay*. Cambridge: Cambridge University Press.
Garthe, C. 2023. *The Sustainable Museum: How Museums Contribute to the Great Transformation*. London: Routledge.
Gregory, J. and Miller, S. 1998. *Science in Public: Communication, Culture and Credibility*. New York: Plenum Press.
Haraway, D. 1984. Teddy Bear Patriarchy: Taxidermy in the Garden of Eden, New York City 1908–1936. *Social Text*, 11: 20–64.
Harrison, R. and Sterling, C. (eds). 2021. *Reimagining Museums for Climate Action*. London: Museums for Climate Action.
Hickman, C., Marks, E., Pihkala, P., Clayton, S., Lewandowski, R.E., Mayall, E.E., Wray, B., Mellor, C. and van Susteren, L. 2021. Climate Anxiety in Children and Young People and Their Beliefs About Government Responses to Climate Change: A Global Survey. *Lancet Planet Health*, 5: e863– e873. https://www.thelancet.com/journals/lanplh/article/PIIS2542-5196(21)00278-3/fulltext
Impey, O. and MacGregor, A. (eds). 1985. *Origins of Museums: The Cabinet of Curiosities in Sixteenth- and Seventeenth-Century Europe*. Oxford: Ashmolean Museum.
IPCC. 2019. *Climate Change and Land*. New York: IPCC. https://www.ipcc.ch/2019/08/08/land-is-a-critical-resource_srccl/
IPCC. 2022. *Fact Sheet Biodiversity: Climate Change Impacts and Risks: Sixth Assessment Report, Working Group II: Impacts, Adaptation and Vulnerability*. New York: IPCC. https://www.ipcc.ch/report/ar6/wg2/downloads/outreach/IPCC_AR6_WGII_FactSheet_Biodiversity.pdf
IPCC. 2023. Summary for Policymakers. In Core Writing Team, Lee, H. and Romero, J. (eds) *Climate Change 2023: Synthesis Report. Contribution of Working Groups I, II and III to the Sixth Assessment Report of the Intergovernmental Panel on Climate Change*. Geneva, Switzerland: IPCC, 1–34.
IPSOS. 2022. *Ipsos Veracity Index 2022*. https://www.ipsos.com/en-uk/ipsos-veracity-index-2022
Janes, R. 2009. *Museums in a Troubled World: Renewal, Irrelevance or Collapse?* New York: Routledge.
Janes, R. forthcoming. *Museums and Societal Collapse*. London: Routledge.
Jowett, P. 2021. £1.1bn Needed to Make UK Theatres Sustainable. *Arts Professional*, 9 November 2011. https://www.artsprofessional.co.uk/news/ps11bn-needed-make-uk-theatres-sustainable
Kolbert, E. 2014. *The Sixth Extinction: An Unnatural History*. London: Bloomsbury.
Krishnan, R., Hidalgo, S.J. and Fuchs, M. 2023. The Problem with Carbon Offsets. *Stanford Social Innovation Review*, Spring 2023: 38–44.
Latouche, S. 2009. *Farewell To Growth*. London: Polity Press.
Lawson, A. 2022. BP Criticised Over Plan to Spend Billions More on Fossil Fuels Than Green Energy. *The Guardian*, 27 December 2022. https://www.theguardian.com/business/2022/dec/27/bp-plan-spend-billions-fossil-fuels-green-energy-oil-gas-renewables

Lewis, G. 1993. Museums in Britain: A Historical Survey. In Thompson, J.M.A (ed) *Manual of Curatorship*. London: Routledge, 22–46.
McGhie, H. 2019. *Museums and the Sustainable Development Goals*. Curating Tomorrow. http://www.curatingtomorrow.co.uk/wp-content/uploads/2020/01/museums-and-the-sustainable-development-goals-2019.pdf
McGhie, H., Mander, S. and Underhill, R. 2018. Engaging People With Climate Change Through Museums. In Filho, W.L., Manolas, E., Azul, A.M., Azeiteiro, U.M. and McGhie, H. (eds) *Handbook of Climate Change Communication* (Vol. 3), 328–348.
Merritt, E. 2019. *TrendsWatch 2019: Truth, Trust, and Fake News*. https://www.aam-us.org/2019/04/17/trendswatch-2019-truth-trust-and-fake-news/
Miller, J.D. 1983. Scientific Literacy: A Conceptual and Empirical Review. *Daedalus*, 11: 29–48.
Museums Association. n.d. https://www.museumsassociation.org/campaigns/museums-for-climate-justice/
NASA. 2022. https://www.nasa.gov/press-release/2021-tied-for-6th-warmest-year-in-continued-trend-nasa-analysis-shows
NASA. 2023. https://climate.nasa.gov/vital-signs/ice-sheets/#:~:text=Key%20Takeaway%3A,adding%20to%20sea%20level%20rise
Neilson, B. and Cameron, F. (eds). 2016. *Climate Change and Museum Futures*. New York: Routledge.
Newell, J., Robin, L. and Wehner, K. (eds). 2017. *Curating the Future: Museums, Communities and Climate Change*. New York: Routledge.
Nisbet, M., Ho, S.S., Markowitz, E., O'Neill, S., Schäfer, M.S. and Thake, J. (eds). 2018. *The Oxford Encyclopedia of Climate Change Communication*. Oxford: Oxford University Press.
Oliver, R. 1971. *Museums and the Environment: A Handbook for Education*. New York: Arkville Press.
Raworth, K. 2017. *Doughnut Economics: Seven Ways to Think Like a 21st-Century Economist*. London: Chelsea Green Publishing.
Rivers, P. 1891. Typological Museums, as Exemplified by the Pitt Rivers Museum at Oxford, and His Provincial Museum at Farnham, Dorset. *Journal of the Royal Society of Arts*, 40: 115–122.
Sloggett, R. and Scott, M. 2023. *Climatic and Environmental Threats to Cultural Heritage*. London: Routledge.
Spence, A. and Pidgeon, N. 2009. Psychology, Climate Change and Sustainable Behaviour. *Environment: Science and Policy for Sustainable Development*, 51(6): 8–18. https://doi.org/10.1080/00139150903337217
Steg, L. and de Groot, J. 2018. *Environmental Psychology: An Introduction*. New York: Wiley-Blackwell.
Sutton, S. (ed). 2022. *The Arts and Humanities on Environmental and Climate Change: Broadening Approaches to Research and Public Engagement*. London: Routledge.
Swyngedouw, E. 2013. The Non-Political Politics of Climate Change. *ACME: An International Journal for Critical Geographies*, 12(1): 1–8.
Thunberg, G. 2022. We Are Not All in the Same Boat. In Thunberg, G. (ed) *The Climate Book*. London: Allen Lane, 154–157.
United Nations. 2021. *Report on Sustainable Development Goal 6*. https://unstats.un.org/sdgs/report/2021/goal-06/

Watson, C. and Schalatek, L. 2020. Climate Finance Thematic Briefing: REDD+ Finance. *Climate Finance Fundamentals 5*, February 2020. https://climatefundsupdate.org/wp-content/uploads/2020/03/CFF5-2019-ENG-DIGITAL.pdf

Watson, P. 2012. *The Great Divide: Nature and Human Nature in the Old World and the New*. New York: Harper.

WHO. 2022. *Ambient (Outdoor) Air Pollution*. https://www.who.int/news-room/fact-sheets/detail/ambient-(outdoor)-air-quality-and-health

WWF. 2019. *WWF Manifesto for Your World*. https://www.wwf.org.uk/sites/default/files/2019-10/election_Manifesto2019.pdf

2

MUSEUMS EMPOWERING CLIMATE ACTION

Sarah Sutton and John Fraser

Introduction/context

The Intergovernmental Panel on Climate Change (IPCC) was established by the United Nations Environment Programme (UNEP) and the World Meteorological Organization (WMO) and then endorsed by the UN General Assembly in 1988. Since that time, it has worked to identify the scientific consensus on the mechanisms that have led to a climate that is projected to no longer be hospitable to the natural systems on which human life depends. Thirty-four years after that group came together, it is clear that this experiment with sharing factual scientific information about the consequences of continued anthropogenic production of atmospheric carbon dioxide and other heat-trapping gasses has failed. It has done virtually nothing to change course for human society.

Two decades before writing this chapter, the United States, where we are based, was navigating a concerted effort to deny scientific fact and political winds that prohibited funding anything even remotely acknowledging the importance of public literacies related to climate change. Though science denial continues, commitment to counteracting its force is expanding. Despite that opposition, science-based museums acknowledged their moral commitment to scientific literacy related to a changing climate. Slowly, but increasingly, arts and humanities museums are doing the same. This work, though, must dramatically expand if the entire sector is to address its moral obligations to nature and society.

Over the first two decades of the 20th century, museum professionals have exhorted their colleagues to advance public engagement with climate change in different settings (Brophy and Wylie 2008; Dorfman 2018; Fraser and Sickler 2009; Janes and Grattan 2019; Janes 2022; Newell et al. 2017; Oliviera et al. 2020; Spitzer et al. 2020). While the tenor of those appeals has moved from potential to

DOI: 10.4324/9781003347606-3

urgency, the narrative is rather consistent. Climate change is well-suited to all types of museums and museum topics, but the number of exhibits and programs on this subject remains much lower than one would expect given the scale of the crisis.

The context for those efforts is changing as the sudden rise in extreme weather volatility; drought; sea level rise; increasing frequency of unprecedented heat, cold, and wind events; and the other catastrophic climate-related crises have made the inevitable obvious. We are now in a position to build on two decades of activist work to overcome the objections of the few remaining denialists to foment change. As a result, the cultural industry, while late to be motivated to embrace their role in action, now has a host of tested precedents and leaders to follow. Therefore, we hope to draw attention to the landmark work now operating below the surface of the status quo in the United States, where climate denial seemed more extreme than in other countries. We do so to illustrate the ease with which the entire museum sector can become a useful tool for thinking about how to migrate from information sharing, to social activation in the waning days of human supremacy over the biosphere.

Distinct value of museums, zoos, gardens, aquariums and historic sites to support this practice

When the authors met in the mid-00s, the United States was roiled by climate denial. We shared our notes and efforts to advance the work in museums, but the majority of public discourse was dominated by doomsday narratives that gravitated around Al Gore's *Inconvenient Truth* initiative, and the Yale Program on Climate Communication was just starting to track American beliefs. Most of the action was driven by the denialist narrative, with science communicators more concerned that the public understood the disaster scenarios being predicted by the science. Which is to say that the concept of climate action at the scale necessary was an elusive idea. While the public knew that recycling toilet paper rolls and driving a bit less might be useful, the scale of solutions and communities working on change were somewhat nascent in society.

But, promise was brewing in the museum sector. Fraser's research at that time demonstrated that the general public was actively interested in zoos and aquariums taking an activist role in recommending climate action (Fraser and Sickler 2009), and the New England Aquarium had begun to invest in studying how to talk about climate change directly with their audiences. That latter effort would blossom through the 2010s into the National Network for Ocean and Climate Change Interpretation, a self-directed volunteer community of practice that can now count more than 40,000 people trained in their techniques and continuing to grow and diversify (Fraser and Switzer 2021).

At this time of writing, there is now empirical evidence that museums are ideally situated to empower climate action. Wilkening's 2022 national study, distributed through communications lists of 186 museums of all types, garnered over

90,000 responses. Her data confirms, without question, that at least 90% think at least some museums should be sharing climate change and climate action and that half of all users responding think the subject matter is appropriate for all museums. Wilkening noted that some museum types forwarding the survey elicited questions about the fit of the topic to the sending museum, but that only explained why these respondents did not claim that all museums should present this information. Despite those caveats, those on museum communication lists rank museums as the second-most trusted source of information on climate, only less than research scientists. Furthermore, 86% of frequent museum-goers expected museums to work to reduce their own carbon footprint. And, as expected, Wilkening's data surfaced the same expected 8.7% of the general public in every survey, the group we can confidently say will never understand or accept that a changing climate is a problem.

So what are the inherent qualities of museum setting and stature, which suit their role as leaders in climate action in their communities?

First and foremost, museums use experience to localize global issues. Today, we are now witnessing our climate future play out as predicted. Heat, cold, and extreme weather events have increased in severity and frequency to the point that every gardener and every person who walks outdoors is aware that things are changing at rates they can describe within their own lifetimes. While a 50-year or 100-year threshold is how climate scientists talk about extreme threat, the museum user can now describe the change in much shorter time cycles.

Second, the museum, as a tool for thinking, is clearly an ideal venue for every type of human experience to situate itself in climate action, and whether city museum or social justice site, memorial museum or botanic garden, there is a climate connection that can use cultural materials to help users think about preparedness and adaptation.

Despite this clear statement of relevance for climate action in museums, some staff remain mired in three swamps of inaction: the false assumption of competitiveness among museums, the false assumption of neutrality, and the fear of losing trust. We can infer from our own experience talking with museum professionals and our data that resistance to taking on the climate as a central topic is a management decision that capitulates to a noisy minority. But this need not be the case.

The first swamp of avoidance is the false notion of competitiveness. Unlike so many other nonprofits in the climate space, museums aren't competitors in their own market. Their type and content offer different ideas and notions that build a more elaborated knowledge in their users. Our studies of the learning ecology in a community demonstrate that the public seeks different types of knowledge on the same topic, and that's why they may visit different types of museums (Gupta et al. 2020). Just because a local science center has a climate change exhibition doesn't mean the zoo and art museum should not do the same. In fact, our studies of zoos and science centers presenting the same information demonstrated a deepened understanding and greater appreciation for a science topic among visitors who saw both exhibitions (Gupta and Plemons 2013). Essentially, we draw on well-established theory that museums are tools for thinking (Carr 2004), they offer different perspectives and philosophies about what can be known about the world, and help use local and global culture to support meaning-making.

Moreover, we return to our own research to assert that museums have a unique affordance over other less trusted sources of information, and distinct from the more trusted research scientists, because they are trusted to be reliable interpreters of knowledge that is often perceived as inaccessible by those who do not feel confident as science thinkers (Dwyer et al. 2020; Fraser and Sickler 2009; Fraser et al. 2008; Rank et al. 2018; Sutton 2020; Sutton and Robinson 2020; Voiklis 2022; Voiklis and Fraser 2022).

The other swamp of avoidance that we believe inhibits museum action is the claim that museums must remain neutral. That clarion call for neutrality is often a simple demand from those in power who do not wish to see a contrary position presented by an organization that is held in high esteem. But the claim that museums are neutral has been thoroughly debunked (see Coffee 2006; Evans et al. 2020; Fleming 2016; Wajid and Minott 2019; Wood 2018). While there may be calls for museums to not challenge prevailing biases because it is discomforting to the comfortable middle classes who patronize them, when a problem becomes self-evident, that public will be quick to turn on the museum and ask why they were not working to bring this issue to the fore at an earlier date.

The third swamp of avoidance that we witness in our work is the fear of losing trust. But trust is not defined by museum users as a single-item scale. Trust is built from five distinct factors that are weighed by their users. Trust is composed of perceived competence, reliability, sincerity, benevolence, and clarity in principles (Voiklis and Fraser 2022). Therefore, to be useful, museums can unpack the fit between their museum's area of inquiry, and what they can present to encourage thinking about our climate future. One need not present the scientific record to activate moral decision-making about what to do to reverse climate change. These moral decisions are the domains of the humanities and well suited to museums of all stripes. There need be no fear of loss of trust as long as the museum doesn't move outside of its principles and competence domains.

Principles that ground climate change communication expectations, needs, and practice

Over the past decade, there's been a plethora of research into effective climate change communications. We've moved well beyond the crisis framing to focus on climate action and climate empowerment. The focus at this writing is anchoring communications at the local level, focusing on how individuals and groups can move their thinking from an abstract issue to a behavioral action or choice (see Bunten and Arvizu 2013; Hoffman 2020; Kretser and Chandler 2020; De Meyer et al. 2020; Fraser and Switzer 2021).

The research behind migrating climate communications from awareness to action should come as no surprise, since they follow the same trajectory as other behavioral interventions, whether they are a 12-step program to overcome addiction or the process people follow to select their mode of transportation. The social theory of learning was clarified more than half a century ago (Bandura 1977). Effectively, humans are drawn to learn in social context, seek to conform to what others they depend on for love and affirmation do, and repeat actions out of habit. It is this process that has led our population into an addiction to the overconsumption of the resources on which all life depends, and to repeat patterns of behavior that accelerate the accumulation of carbon in the atmosphere that will be our own demise.

Since the 1960s, Prochaska has elaborated on his transtheoretical model for behavior change that can be simply characterized as a period when there is lack of awareness of a need to change or a pre-contemplation stage (meaning someone doesn't have enough information to think about something), a contemplation stage when they are aware of a need to change, but not necessarily convinced themselves to take action, followed by experimentation through action until that change becomes a habit (Prochaska 2020).

While we could attribute Prochaska's model to museum visitors whom we target for driving to take action, for this chapter, we believe that readers at this point in their reading are likely in the same domain as the visitors, so we will use the remainder of this chapter to help those contemplating change see that the paths to museum action for a better climate future are proven, established, and easy to implement. We have five examples of climate change exhibits in design or installed that call visitors to action. Each relies on two or more of the trust factors identified by Voiklis and Fraser (2022): perceived competence, reliability, sincerity, benevolence, and clarity in principles. Their examples are part of this new era of climate change engagement with the public.

Mount Desert Island Historical Society

Landscape of Change is a multi-partner, multi-modal climate change engagement effort originating with the Mount Desert Island Historic Society (MDIHS) in Maine. The approach uses the island experience to localize global issues, focusing on the trust factors of Reliability (using both historical collections and scientific study) and Sincerity (basing the work in the conservation history and ethic of the island) (author interview, 16 August 2022).

This year-round community has a significant summer-home and vacationer population, but both publics value this landscape. MDIHS Executive Director Raney Bench has observed climate change and its impacts on the island over the last 15 years and sees the biggest problem around climate change as the absence of a shared understanding of what's happening to the island and its communities. Where there is no agreement on the problem or its solutions, there can be no progress. Bench knows that the story of Mount Desert Island, as illustrated in the MDIHS collection, can tell a climate story without politics but something just as powerful – the understanding of change in the place they all love. Bench is using measurements, historic and present-day, she surfaces from the collection and other local records to help the community begin to articulate what it needs to do. Her focus is on active engagement with what the community sees happening, capitalizing on local experience to bring understanding to the global climate crisis. She partnered with the scientists at Schoodic Institute and subsequently with artists, Acadia National Park, College of the Atlantic, the MDI Biological Laboratory, and A Climate to Thrive (a nonprofit strengthening sustainable choices by residents). The MDIHS work has fostered scientific and humanities research that supports exhibits and public engagement experiences, a museum without walls. Using art, science, humanities, and conversation, MDIHS is demonstrating a changing climate in ways that any visitor, and any summer or year-round resident, can value and embrace.

This is a series of summer exhibits, each examining the history of science and conservation on the island, using citizen science and cultural history to engage the public in the research on climate variation at MDI and the vulnerabilities of the island to climate change. Year one examined birds, pollinators, seawater and temperature levels, mining historical scientific data, and cultural collections to compare to current data and for use in anticipating future scenarios. Year two shared six original works of art by Jenn Steen Booher representing her observed impacts of climate change on the development of oyster landings (seed oyster anchor sites shifting with warming waters); alewife spring migration die-off; crash of shrimp populations and the rise of squid; softshell clam populations diminished; and the rise of green crabs and ribbon worms. The banners are installed in the lobbies of the local theater and bank, but sometimes copies travel to events around the island to outdoor spaces. The images are highly engaging, with charts-as-art, or art-as-charts, blending information on pollinators and bird population data. There is a different scientist at every location to discuss what's happening in that landscape.

For 2023, the focus is sea level rise and king tides. The team is working with high school students to identify vulnerable places on the island; map historic sea levels; and to observe current king tide events and map those. A public art piece will capture the discoveries. Related programs had pre-teens helping to design an art installation in a few marshes at historic height of tide and projected in the most vulnerable spots, creating a museum without walls focused on sea level rise. MDIHS has just received a highly competitive grant award to work with high school students to review primary source records for the height of tide recordings by ship captains and harbor masters and to find collection records of king tides and storm surge measurements in newspapers, diaries, and business records. For modern comparisons, they'll be out on the landscape during these events to take measurements and share the experience and information with next summer's visitors.

Smithsonian Institution Traveling Exhibition Service

Carol Bossert is Program Manager for Science, Smithsonian Institution Traveling Exhibition Service (SITES). She leads the work to create *Knowing Nature: Stories of the Boreal Forest,* a SITES traveling exhibit opening in Spring 2023, at the MSU Museum in East Lansing, Michigan, a Smithsonian Affiliate. The approach uses the physical experience of the exhibit space to evoke the distant landscape affecting and affected by global climate change. The process highlights the trust factors of Perceived Competence (the scientists and the Smithsonian Institution name), Reliability (again the Smithsonian Institution name as a credible source for both the borrowing institution and the visitors to the exhibit), and Benevolence (imparting climate communication skills to the museum staff and illustrating the commitment of Indigenous people to caring for the boreal forest for humanity, not just themselves) (author interview, 31 August 2022).

The exhibit has its genesis in the work of scientists at the Smithsonian's National Museum of Natural History (NMNH). They have long felt that the boreal forest in the Arctic is 'probably one of the most important ecosystems that no one ever heard of,' Bossert says, and that while the tropical rainforests have been a hotspot of interest, the northern forest is a cold spot for public attention; yet, it's a carbon 'bomb' waiting to go off if public commitment to slowing global warming fails to prevent the loss of this massive carbon sink for the planet. But because climate communication research has demonstrated that abstract ideas and statistics do not motivate humans to change their behaviors (Corner et al. 2018), the team wanted to illustrate the real-world connection of the forest to individual health, how every collective and individual action of someone in the United States is impacting that northern forest, and why the forest is so important to climate change and our future. To create an exhibit experience driving attention to this topic in ways that could build connections, and perhaps lead visitors to take action, requires emotional engagement that could only be driven by stories more humanistic than data based.

Bossert and her team were also committed to sharing the messages from the project's Indigenous advisors that everything that people do to protect the forest and build awareness of its role in nature and the health of the planet they do because of their knowledge that we are all part of the land, not separate from it. As they do this work, they are modeling ways that we can protect our local lands no matter where we live.

That benevolence aspect, on a much smaller scale, is evident in Bossert and her team's effort to ensure that the exhibit experience leaves behind resources for the staff at museums borrowing the exhibit: training on climate change and climate communication that they can use long after the 12-week loan is complete.

This leave-behind process benefits the institution over the long term that offsets the difficulty, Bossert says, of setting aside the time to train staff for such a short exhibit run.

> If we can create a resource they will use, as a way of expanding their expertise in talking about climate change, that is something they can use for their permanent gallery, and on climate change work far into the future.

Natural History Museum of Utah

Lisa Thompson is an Exhibit Developer at the Natural History Museum of Utah, developing an exhibit currently called *A Climate of Hope*, with an anticipated opening of Fall 2023. The exhibit will replace a much smaller, very traditional climate change exhibit installed as part of a larger gallery in the brand new museum in 2011. The approach uses the Utah experience and culture to localize global issues and optimizes the exhibit as a tool for thinking. It focuses on the trust factors of Reliability (it is the State's natural history museum) and Clarity in Principles (basing the work in the cultural values of community and preparedness) (author interview, 1 September 2022).

The original exhibits demonstrated what many in the field still felt was the best approach: the hockey stick graph and data-driven message. At the time, it was 'bold for Utah in that political climate,' Thompson says. Knowing what we do now, 'visitors then may have struggled to see trends in the spikey graph of change over centuries, and to process the data driven, fact-based message' in the small gallery.

The design team's deep dive into climate communication research led them to an approach that makes climate change local, and tangible, and to encourage visitors to move beyond necessary individual actions to necessary collective and collaborative actions. Among the materials that contributed to their thinking was Hayhoe's approach to rational hope (2021). Their study made it clear that this would not be a climate science exhibit but based on the science of climate change communication and one that created for visitors 'the possibility of experiencing rational hope.' The exhibit uses frameworks from both Eco America and the National Network for Ocean and Climate Change Interpretation (NNOCCI) and a solid grounding in Yale Climate Communications resources.[1]

The exhibit begins as the visitor enters an Aspen grove, important as a much-loved state tree in the Utah landscape and one threatened by climate change. The circular space does more than create a feeling of awe and magic – it illustrates how a grove is clonal, that the grove is a single organism: the trees may appear to be individuals, but they are all connected. 'It's a metaphor for community,' Thompson says, 'it seems that we are separate individuals, but we and they are all connected and threatened by climate change.'

Quickly the exhibit turns to emphasizing the local, relevant, and concrete aspects of climate change, with a brief look at local impacts, such as how people experience climate change in their daily lives. As an example, the commonly expressed experience of shorter and warmer winters is easily illustrated by historic photos of skaters on the familiar pond in Liberty Park. Some may remember skating there regularly, but now it rarely freezes over. Visitors will be asked to share what they're experiencing and how it's impacting their lives. The purpose is to examine what climate change 'means for us, and what do we do about it?' Rather than focusing on the crisis, the emphasis is on preparedness, an important part of Utah culture. The prototyping process for the diorama on what a climate-adapted Utah looks like revealed that many visitors were not aware of adaptations already in place. The team is adding contemporary images such as the library and the convention center doubling as cooling stations.

Just as preparedness is an important part of Utah culture and a pathway to climate conversations, air quality is a topic that will also reach a broad group of visitors. Salt Lake City's air conditions are frequently out of compliance with health standards. Summer ozone and winter inversions trapping particulates are already common experiences, and climate is driving wildfire smoke pollution. The exhibit illustrates how actions that Utah's citizens can take to address air quality are the same for addressing climate change. To build momentum, the focus shifts to solutions already underway in Utah and all the ways people are taking action. Since their prototyping indicated that many visitors didn't have a clear understanding of what a renewable energy transition would look like, they'll create a map of utility scale energy projects already happening in Utah with the message that 'we've always been an energy state, and we can be a renewable energy state with our incredible resources.' Next to the map will be a video paired with a diorama of a community that walks visitors through the transition to renewable electricity. The narrative and images will emphasize that the transition is already underway due to the plummeting prices of solar and wind energy and that it will have many benefits, including lower energy bills, clean the air, *and* helping address climate change.

The conclusion shifts to the visitors: what they know, what they can do, what will make a difference. Though visitors participating in prototyping wanted a checklist of ten things to do, the team resisted this default and mistaken response and instead focused on encouraging many more conversations about climate change (rather than debates), asking visitors to take actions a level above individual action. The team found that the word 'collective' action was more of a block than a gate and instead that 'community level action' really resonated. Prototyping landed them in a two-part experience of simple interactives. The first helps visitors understand the difference in impact between individual- and community-level actions when the goal is systems' change. Visitors are invited to throw rocks labeled with different climate actions into a digital pond to makes waves big enough to change our energy system. Individual actions create small ripples; community-level actions create larger waves and trigger other rocks to be thrown into the pond – as if another group heard about your group's project to electrify school buses, for example, and copied it. After a visitor throws several community-level action rocks into the pond, a big wave washes across the screen, and the energy system transition advances.

After being primed with this activity to shift thinking from individual-level to community-level actions, visitors will be invited to explore the community-level actions they could take at a station for creating a personal Venn diagram of activities. Visitors can fill three circles representing (1) what they are good at or interested in, (2) what groups they are part of or would like to join, and (3) the challenges facing their community. Visitors testing the prototype enjoyed sorting their options into the diagram, and their comments revealed what they had learned from exhibit components. Thompson adds that though no one will find a clear path in their diagram, they can find 'a process to the path for action.'

Smithsonian Institution's National Museum of Natural History

Siobhan Starrs is an Exhibition Developer and Project Manager at the Smithsonian Institution's National Museum of Natural History (NMNH), where she helped develop the permanent exhibition *Deep Time: The Hall of Fossils* (2019), the first fossil exhibition in the world to embed climate change so overtly. Deep Time's main message reinforces that the study of the past (paleobiology, paleoecology, and paleoclimatology) informs our understanding of humanity's role in shaping the present and the future. In 2022, she and colleagues at NMNH developed a small exhibit, *Our Places: Connecting People and Nature*, as a testing space for the museum's next major installation – a 10,000-square-foot long-term installation called the *People In Nature Experience*. Our Places tests approaches for localizing a global issue, introducing environmental justice concepts, and optimizing exhibit spaces as a tool for thinking and community engagement. These projects focus on the trust factors of Reliability (it is the nation's natural history museum), Perceived Competence (it's the Smithsonian Institution), and professional Benevolence (creating a multi-institution sharing group on climate change exhibit development).

The seed exhibit's place-based approach was based on feedback from visitors and climate change communication researchers. The goals were to (1) connect people to nature: next-door nature and capital 'N' nature; (2) illustrate how place matters to people: because so many people are disconnected or report feeling disconnected from nature, the goal was to marry people and place, to return that connection, despite its indoor setting; and (3) provide a call (and a path) to action. The small but highly participatory gallery space was a new approach for the nation's natural history museum: instead of a national story, it was a place-based, *this* neighborhood-place based, exhibit to help the staff understand how and what local communities, not just traveling visitors, will visit, engage, and expect to discover. The team for *Our Places* wondered how to inspire people to love, protect, and act. The path was to connect people to a place, creating a core of empathy in that place at least for the visitor but also for the place and for others who value that space. They used the activity spaces in the gallery to shift from awe to connection, from care to empathy, to action (author interview, 2 September 2022).

As Starrs and colleagues concluded the Deep Time development work in 2017, they were thinking 'what next?'. Starrs reached out to colleagues at other natural history museums to create a monthly online video meeting to discuss their current work. The gathering is now in its third year. The members are sharing climate change exhibit concepts, designs, and, soon, results that help each complete their work more confidently and feel more supported and improve the production process and public impact of each other's exhibits. Thompson and Bossert, mentioned earlier, and Jen Kretser and Stephanie Ratcliffe, introduced later in the chapter, are part of that learning group.

The Wild Center

The Wild Center (TWC) is known for its climate programming, particularly its Youth Climate Summit, rather than climate exhibit work. For a long time, the Director, Stephanie Ratcliffe, was uninspired by the extreme effort to create an exhibit that simply named the problem or tried to persuade visitors with scientific knowledge alone into scientific understanding. But when Jen Kretser, Director Climate Programs, suggested that TWC use Project Drawdown's Framework,[2] Ratcliffe could see a solution-framing focus that made sense to her, a concrete and cohesive strategy for moving people forward in climate action. The exhibit localizes global climate change, and the approach uses the trust factors of Sincerity (commitment to experience of the Adirondack communities), Reliability (the examples of people from the community and TWC's record of valuable climate action), and professional Benevolence (encouraging the copying of these ideas) (author interview, 8 September 2022).

Ratcliffe tells the story of a turning point moment for her in addressing climate change in her institutions. She says she left a discussion among peers who are part of the Association of Science-Technology Centers, knowing in her bones that it's no longer about sharing the facts about climate change, or making the case to convince anyone, it's about sharing solutions – for individuals and museums. The core of the exhibit are the stories of real people doing this good solutions work right now.

The staff identified the exhibit audience using the Yale Six Americas values segmentation instruments (Maibach et al. 2011). Their visitor and member polls showed that over 80% were in the 'Concerned' and 'Alarmed' categories, not the middle as expected. Then the team worked with this group for online prototyping of exhibit components (due to COVID) and focus group conversations. Those discoveries, combined with climate communication practices learned from the NNOCCI training, boosted team confidence in a solutions-based, people-focused approach. The planning required a reexamination of the institution's just-the-facts approach to exhibit work and shifted them to clearly offering productive solutions that are reliable options for self-directed choices for visitors.

The TWC team created a matrix of solutions appropriate for the region and showcased individuals who were deploying these solutions in the region at the time. Their stories *are* the exhibit. Ratcliffe is sure that for future climate exhibits, solutions framing and storytelling approaches will remain priority pathways. And the team will continue their learning journey on the sustainable materials in physical museum design.

Ratcliffe and Krester encourage others to use existing material and research to do their work, as TWC used NNOCCI, Project Drawdown, and Yale Climate Communications to do theirs and to give away their discoveries. 'This is not the time for selling [approaches and solutions to climate change]; rather it's time [for museums] to figure out how to give our intellectual property to each other for free.' She calls it The Big Giveaway. 'We want people to steal our stuff, and use it; we want to give it away.' 'We all need to do [this work] and as fast as we can if we're going to be catalysts in our communities.' 'Don't do new; just steal and run with it.'

NNOCCI: a museum community of practice for climate action

The museum industry has a long history of thinking of each museum as a community of autonomous actors working independently with their service populations, and their associations as connectors and advocacy entities. But museums are not exactly unique, and each type of museum tends to pursue remarkably similar approaches, content, and delivery as their kin. Science centers have a history of designing and selling traveling exhibits to one another; zoos and aquariums freely share their ideas so others can replicate, adapt, or copy materials or strategies. And many of the major museum design firms draw on a proven kit of parts.

What we know from consumer marketing is that repetition is essential for consumer awareness. And, as we noted earlier with Gupta and colleagues' work (Gupta et al. 2020), the people who use museums tend to visit more than one to advance their learning. So it stands to reason that climate communications at museums will have greater impact if they are found in many of the museums people visit.

In 2007, Billy Spitzer, the Vice-President for Education at the New England Aquarium, reached out to the Frameworks Institute, a group that studies how people think and talk about social issues to develop recommendations for communications that can spark change. Billy proposed that climate change and ocean health were two complex issues that needed a more careful study to align to how aquariums talk to their publics about the issue. With funding from the Institute of Museum and Library Services, they developed new principles for climate communication based on persistence-tested framing language (Bunten and Arvizu 2013) and trauma-informed training (Swim and Fraser 2014).

Based on the success of that work, and convergent with John and his colleagues' (Fraser et al. 2013; Fraser and Brandt 2013) initial studies demonstrating the debilitating emotional consequences educators and conservationists were experiencing as they dealt with the obvious evidence of climate change, they proposed a two-pronged concept for disseminating their communications strategy. Together, the teams proposed a training program based on using the tested communications tools that were showing promise and a trauma-informed, emotionally present training that was aimed at building a Community of Practice (CoP) dedicated to ongoing growth and expansion of the tools. The result was the foundation of the National Network for Ocean and Climate Change Interpretation. From an initial two cohorts of ten institutions each, sending two participants to the training, today the community has grown to represent more than 40,000 trainees and is a self-governing completely volunteer CoP that continues to develop new techniques, applications, and content that they share with the network.

NNOCCI was also committed to careful research into the practice and principles behind the program. The results substantially advanced the understanding of how hope and self-efficacy were central to advancing both successful communications (Geiger et al. 2017a, 2017b, 2019, 2021; Swim and Fraser 2013, 2014; Swim et al. 2014, 2017, 2018). It confirmed that social networks with many museums using the same messaging and providing support for one another could have direct impact on public attitudes and voters in particular. Early results also revealed that museum visitors are more likely to be active voters, were more likely to talk to others about voting decisions, and turned to the museums that present that information for guidance in their advocacy.

The NNOCCI experiments revealed that the primary limitation on climate communication was self-editing by those charged with the task. The emotional burden of thinking about the issue, concerns about not fully understanding the science, and fear of denialists all limited willingness to present the content, even if it was part of their job. After trying the new techniques, receiving positive feedback, and knowing you were part of a community working across the country for common goals, all allayed the prior fears and resulted in substantially more action. Quite simply, fear prevented doing something that will be well received if you just give it a try. And supporting hope was a central mechanism to start a new conversation with people willing to advocate for climate action.

Fifteen years after founding NNOCCI, we now have unassailable evidence that the museums are ideally suited to be catalysts for engaging the public in climate action (Fraser and Switzer 2021, Ch. 11). They no longer need to constrain themselves to passive observation and can now consider how many ways they can collaborate for more effective social change. The NNOCCI.org is now a clearing house for new resources and techniques that are open access and available for use by any museum.

The NNOCCI model can be considered a psychological response to environmentalists being called snowflakes. The model demonstrates that one action alone will not cause change, but when enough of us get together and do things the same way, we are a snowstorm, capable of changing the behavior of everyone in our communities.

Today, the literature supports the community of practice approach that bridges museum type and practice. The community has developed assets for visualizing the change and fostering local dialogue that addresses the unique conditions (see Geiger, Swim and colleagues).

At this writing, the social sciences have moved from activating dialogue to direct calls for climate action. As reports from a cross section of the infrastructures that define contemporary society have shown, there is no part of the economy or society that is exempt from entering a transition toward a regenerative ecology (see De Meyer et al. 2020; Niepold et al. 2022 for examples). The dialogues about climate action are no longer an issue of understanding the problem, and most museums can skip the climate awareness and focus on their role in activating action in all aspects of their work, from operations, communications, exhibition, and, most importantly, working with their communities to implement solutions wherever they are based.

Conclusion and necessary steps to significantly expand this practice in the sector

As this chapter has shown, the call for pilot studies and testing of techniques has passed. The models are obvious, and visitors expect to see museums standing up for climate action. We suggest that the problem is a belief in museum exceptionalism that every museum must recreate the wheel in order to do something. And we suggest that nothing could be further from the truth. Instead, we now call on museums to simply copy what's working and add their voice to the chorus about the need for climate action.

Museum users are demanding that museums restructure operations and physical buildings as models of climate action. Museum users are also demanding that museums take a position on climate as a human right and imperative. And museum users know that other museums are already doing it, so they will not look favorably on the museums that continue to resist action.

We close this chapter with some hope that the recalcitrance in the field may soon be overcome. Recent reports are moving well beyond what an individual museum can do and expect all museums to find their place as part of the chorus demanding that climate action is a social and environmental justice issue that can no longer be ignored.

Notes

1 https://climatecommunication.yale.edu/
2 https://drawdown.org/drawdown-framework

References

Bandura, A. 1977. Self-Efficacy: Toward a Unifying Theory of Behavioral Change. *Psychological Review,* 84(2): 191–215.

Brophy, S. and Wylie, E. 2008. *The Green Museum: A Primer on Environmental Practice*. Lanham, MD: Rowman and Littlefield.

Bunten, A. and Arvizu, S. 2013. Turning Visitors Into Citizens: Using Social Science for Civic Engagement in Informal Science Education Centers. *Journal of Museum Education*, 38(3): 260–272.

Carr, D. 2004. Reading Beyond the Museum. *Journal of Museum Education,* 29(1): 3–8.

Coffee, K. 2006. Museums and the Agency of Ideology: Three Recent Examples. *Curator: The Museum Journal*, 49(4): 435–448.

Corner, A., Shaw, C. and Clarke, J. 2018. *Principles for Effective Communication and Public Engagement on Climate Change: A Handbook for IPCC Authors*. Oxford: Climate Outreach.

De Meyer, K., Coren, E., McCaffrey, M. and Slean, C. 2020. Transforming the Stories We Tell About Climate Change: From 'Issue' to 'Action'. *Environmental Research Letters*, 16(1): 015002.

Dorfman, E. (ed). 2018. *The Future of Natural History Museums*. London: Routledge.

Dwyer, J.T., Fraser, J., Voiklis, J. and Thomas, U.G. 2020. Individual-Level Variability Among Trust Criteria Relevant to Zoos and Aquariums. *Zoo Biology*, 39(5): 297–303. https://doi.org/10.1002/zoo.21562

Evans, H.J., Nicolaisen, L., Tougaard, S. and Achiam, M. 2020. Perspective: Museums Beyond Neutrality. *Nordisk Museologi*, 29(2): 9–19.

Fleming, D. 2016. Do Museums Change Lives? Ninth Stephen Weil Memorial Lecture. *Curator: The Museum Journal*, 59(2): 73–79.

Fraser, J. and Brandt, C. 2013. The Emotional Life of the Environmental Educator. In Krasney, M. and Dillon, J. (eds) *Trading Zones: Creating Trans-Disciplinary Dialogue in Environmental Education.* New York: Peter Lang, 133–158.

Fraser, J., Pantesco, V., Plemons, K., Gupta, R. and Rank, S.J. 2013. Sustaining the Conservationist. *Ecopsychology*, 5(2): 70–79.

Fraser, J. and Sickler, J. 2009. *Why Zoos and Aquariums Matter: Handbook of Research Key Findings and Results From National Audience Surveys*. Barcelona, Spain: Association of Zoos and Aquariums.

Fraser, J. and Switzer, T.F. 2021. *The Social Value of Zoos*. Cambridge: Cambridge University Press.

Fraser, J., Taylor, A., Johnson, E. and Sickler, J. 2008. The Relative Credibility of Zoo-Affiliated Spokespeople for Delivering Conservation Messages. *Curator: The Museum Journal*, 51(4): 407–418.

Geiger, N., Gasper, K., Swim, J.K. and Fraser, J. 2019. Untangling the Components of Hope: Increasing Pathways (Not Agency) Explains the Success of an Intervention that Increases Educators' Climate Change Discussions. *Journal of Environmental Psychology*, 66. https://doi.org/10.1016/j.jenvp.2019.101366

Geiger, N., Gasper, K., Swim, J.K., Fraser, J. and Flinner, K. 2021. How Do I Feel When I Think About Taking Action? Hope and Boredom, Not Anxiety and Helplessness, Predict Intentions to Take Climate Action. *Journal of Environmental Psychology*, 76. https://doi.org/10.1016/j.jenvp.2021.101649

Geiger, N., Swim, J.K. and Fraser, J. 2017a. Creating a Climate for Change: Interventions, Efficacy and Public Discussion About Climate Change. *Journal of Environmental Psychology*, 51: 104–116.

Geiger, N., Swim, J.K., Fraser, J. and Flinner, K. 2017b. Catalyzing Public Engagement With Climate Change Through Informal Science Centers. *Science Communication*, 39(2): 221–249.

Gupta, R. and Plemons, K. 2013. Volunteer Perceptions of Public Value From a Co-Hosted Museum-Zoo Exhibition. *Informal Learning Review*, January-February: 11–15.

Gupta, R., Voiklis, J., Rank, S.J., Dwyer, J.T., Fraser, J., Flinner, K. and Nock, K. 2020. Public Perceptions of the STEM Learning Ecology – Perspectives From a National Sample in the US. *International Journal of Science Education, Part B*, 10(2): 112–126. https://doi.org/10.1080/21548455.2020.1719291

Hayhoe, K. 2021. *Saving Us: A Climate Scientist's Case for Hope and Healing in a Divided World*. New York: Simon and Schuster.

Hoffman, J.S. 2020. Learn, Prepare, Act: 'Throwing Shade on Climate Change'. *Journal of Museum Education*, 45(1): 28–41.

Janes, R.R. 2022. The Value of Museums in Averting Societal Collapse. *Curator: The Museum Journal*, 65(4): 729–745.

Janes, R.R. and Grattan, N. 2019. Museums Confront the Climate Challenge. *Curator: The Museum Journal*, 62(2): 97–103.

Kretser, J. and Chandler, K. 2020. Convening Young Leaders for Climate Resilience. *Journal of Museum Education*, 45(1): 52–63.

Maibach, E.W., Leiserowitz, A., Roser-Renouf, C. and Mertz, C.K. 2011. Identifying Like-Minded Audiences for Global Warming Public Engagement Campaigns: An Audience Segmentation Analysis and Tool Development. *PLoS One*, 6(3): e17571.

Newell, J., Robin, L. and Wehner, K. (eds). 2017. *Curating the Future: Museums, Communities and Climate Change*. Lanham, MD: Rowman and Littlefield.

Niepold, F., Fraser, J., Crim, H. and Field, S. (eds). 2022. *A Proposed Climate Empowerment Theory of Change for the United States of America, Its Territories, Protectorates and the Sovereign Nations within Its Borders*. Knology. https://bit.ly/ClimateEmpowerment_ToC

Oliviera, G., Dorfman, E., Kramar, N., Mendenhall, C.D. and Heller, N.E. 2020. The Anthropocene in Natural History Museums: A Productive Lens of Engagement. *Curator: The Museum Journal*, 63(3): 333–351.

Prochaska, J.O. 2020. Transtheoretical Model of Behavior Change. *Encyclopedia of Behavioral Medicine*, 2266–2270.

Rank, S.J., Voiklis, J., Gupta, R., Fraser, J. and Flinner, K. 2018. Understanding Organizational Trust of Zoos and Aquariums. In Hunt, K.K. (ed) *Understanding the Role of Trust and Credibility in Science Communication: Proceedings of the Iowa State University Symposium on Science Communication*. Ames, IA. https://lib.dr.iastate.edu/sciencecommunication/2018

Spitzer, W., Fraser, J., Sweetland, J. and Voiklis, J. 2020. Applied Social Science to Scale Climate Communications Impact. In Henderson J. and Drewes, A. (eds) *Teaching Climate Change in the United States*. London: Routledge, 123–142.

Sutton, S. 2020. The Evolving Responsibility of Museum Work in the Time of Climate Change. *Museum Management and Curatorship*, 35(6): 618–635.

Sutton, S. and Robinson C. (eds). 2020. The Climate is Changing: Why aren't Museums? *Journal of Museum Education*, 45(1).

Swim, J.K. and Fraser, J. 2013. Fostering Hope in Climate Change Educators. *Journal of Museum Education*, 38(3): 286–297.

Swim, J.K. and Fraser, J. 2014. Zoo and Aquarium Professionals' Concerns and Confidence about Climate Change Education. *Journal of Geoscience Education*, 62(3): 495–501.

Swim, J.K., Fraser, J. and Geiger, N. 2014. Teaching the Choir to Sing, Use of Social Science Information to Promote Public Discourse on Climate Change. *Journal of Land Use and Environmental Law*, 30(1): 91–117.

Swim, J.K., Geiger, N., Fraser, J. and Pletcher N. 2017. Climate Change Education at Nature-Based Museums. *Curator: The Museum Journal*, 60(1): 101–119.

Swim, J.K., Geiger, N., Sweetland, J. and Fraser, J. 2018. Social Construction of Scientifically Grounded Climate Change Discussions. In Clayton, S. and Manning, C. (eds) *Psychology and Climate Change: Human Perceptions, Impacts, and Responses*. Amsterdam, Netherlands: Elsevier Inc., 65–93.

Voiklis, J. 2022. Key Concepts: Trust. *ACM Trends*, 5(3): 1–4.

Voiklis, J. and Fraser, J. 2022. *What Does It Mean to Trust a Museum? AAM Center for the Future of Museums*, 3 March 2022. https://www.aam-us.org/2022/03/01/what-does-it-mean-to-trust-a-museum/

Wajid, S. and Minott, R. 2019. Detoxing and Decolonising Museums. In Janes, R.R. and Sandell, R. (eds) *Museum Activism*. London: Routledge, 25–35.

Wilkening, S. 2022. *Annual Survey of Museum-Goers 2022*. www.wilkeningconsulting.com

Wood, C. 2018. *Visitor Trust When Museums are Not Neutral* (Doctoral dissertation).

3

THE EMERGENCY IS AN OCTOPUS! MUSEUMS ACTIVATING PUBLIC IN A PLANETARY EMERGENCY

Bridget McKenzie and Victoria Burns

Introduction

This is an account of experiments by Climate Museum UK (CMUK) and other museums aiming to activate people in response to the Earth crisis. As a distributed museum, our practitioners carry out 'activations' in their communities across the UK, growing collections through local participatory projects with partner organisations designed to take people on a journey through feeling and thinking, towards taking effective action.

With such activist practice, neutrality is impossible because decarbonisation and political violence are entangled, locked into the historical roots and continuing power of bio-colonialism. The chances of breaching 1.5°C of global warming by 2025 have doubled since a similar assessment in 2021 (Galey 2021), and an IPBES/IPCC report has warned that biodiversity loss worsens the crisis (Pörtner et al. 2021). The worse the reported crisis becomes, and as environmental, economic and social systems begin to collapse, the more our personal and collective defence mechanisms kick in. These mechanisms include denial (failing to accept facts), disavowal (failing to make personal sense of facts) and rationalising (jumping to single solutions) (Lertzman 2009). The psychological and social challenges of anticipating and responding to the impacts of our disrupted biosphere are immense.

This chapter describes the principles and practices that are shared between the associates and partners of Climate Museum UK, which seek to address these interpersonal challenges in ways that go beyond therapeutic amelioration to enable transformations in audiences and museum teams that maximise the network effect of activations for systemic change.

DOI: 10.4324/9781003347606-4

The emergency is an octopus!

The increasingly troublesome context underpins our call for radical activation for systemic change. This emergency can be seen as one situation: an Earth crisis of breached planetary boundaries leading to existential risks and uncertainties, encompassing every other issue. Alternatively, it can be seen as multiple intersecting emergencies for humans and other species, including conflict, inequalities, displacement and pandemics. Most typically, it is described only in relation to climate, with this used as a synecdoche for the wider Earth crisis. Our activations with cultural professionals seek to broaden this framing, suggesting that museums should be precise with terminology because we play a trusted public role as educators and should be distinctively capable of illuminating complex systems.

Moreover, if the focus is tightly on technical solutions and carbon emissions, this can exclude perspectives of nature-connected people and more-than-human beings. Indeed, overpromising technological solutions diminish the urgency surrounding the imperative to end the extraction of fossil fuels (Warszawski et al. 2021) and make reparations for their harm. The promotion of 'Net Zero' is driven by a need to protect the status quo and avoid systemic change. It is part of mindsets that circumvent difficult truths about our colonial history and the profound cultural changes demanded to address the Earth crisis. In CMUK, we use tools such as our Earth Crisis Blinkers graphic (Figure 3.1) to help audiences expand their scope. In removing the blinkers to see the whole Earth in crisis, we can see causes, impacts and solutions in a less linear way. Some factors that might be ignored, such as indigenous power and racial justice, are significant levers for just and ecocentric system change.

In May 2022, the United Nations published its 'Global Assessment Report on Disaster Risk Reduction', which points to escalating intersections between disasters, vulnerabilities of economies and failures of ecosystems, risking global civilisation collapse. Its abstract highlights that 'the best defence against future shocks is to transform systems now, to build resilience by addressing climate change and to reduce the vulnerability, exposure and inequality that drive disasters' (Global Assessment Report on Disaster Risk Reduction 2022). Museums can contribute to this defence by being agents for the transformation of social, economic and ecological systems, if there is a groundswell of recognition among our profession about the scale and dimensions of the Earth crisis and an expanded understanding of the roles of Culture to respond.

James Lovelock, whose notion of Gaia has kept many environmentalists optimistic about the regenerative potential of Earth, said, not long before his death in July 2022,

> Up until now, the Earth system has always kept things cool on the Earth, fit for life, that is the essence of Gaia. It is an engineering job, and it has been well done. But I would say the biosphere and I are in the last 1% of our lives.
>
> *(Watts 2020)*

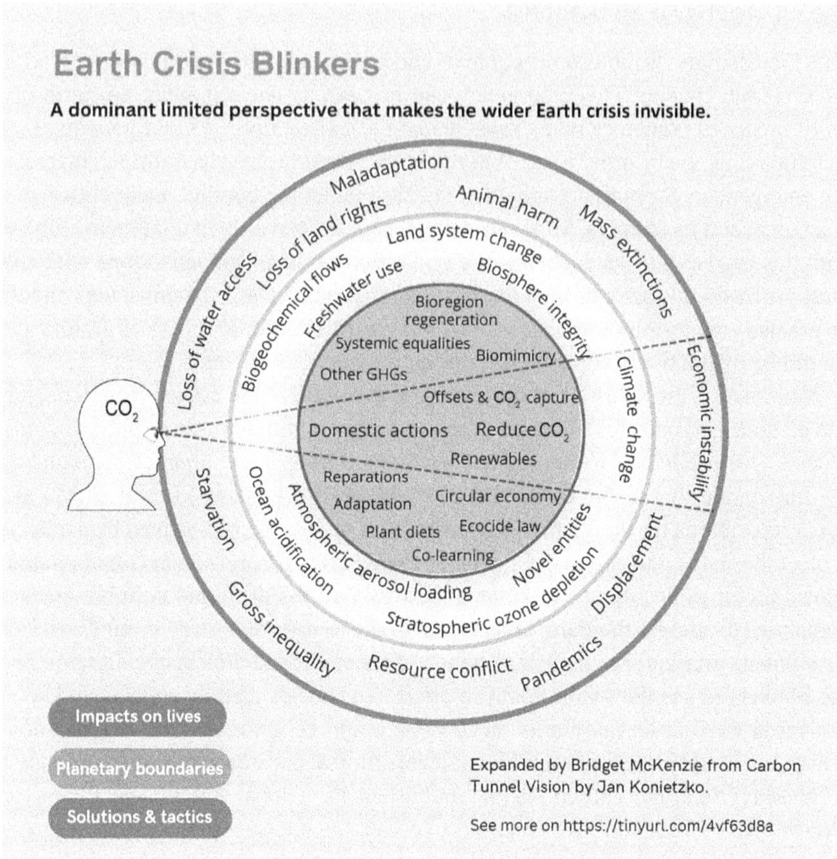

FIGURE 3.1 Diagram of Earth Crisis Blinkers used in CMUK training

The catastrophe often imagined to come in some indeterminate time, for example, in our grandchildren's lifetimes, is happening now. This octopus-like emergency has grown extremely large, very quickly, and is now negatively affecting all areas of the planet, at all latitudes and all levels of advantage. Alex Steffen has called this the Transapocalypse (Steffen 2021) whereby catastrophic events are hitting privileged regions of the planet not expected to be impacted for decades to come. Climate science in 2022 brings more alarms than ever before: the Antarctic ice sheet is losing twice as much ice as previously thought (Stokes et al. 2022); there has been a six-fold increase in famine conditions over the past two years (Oxfam International 2021) and, as we write, a third of Pakistan is under water. The public in the Global North, who make up the majority of tourists and museum visitors, are now directly affected and deeply concerned. A poll in 2019 found that more than half of respondents in France, Italy, the United Kingdom and the United

States think civilisation as they know it will collapse in years to come (Cassely and Fourquet 2020). This data supports the view that planetary systems, including human civilisation, are already collapsing if we think more in more granular and nuanced ways about what collapse is. Some places have already experienced collapse of ecosystems and societies, increasing with extractive colonialism over centuries. This has joined into a patchwork of collapse, currently overlaid with the hugely disruptive factor of global warming.

Strategic advocacy for effective activation

As a team in CMUK, we work at a grounded level of local relationships with communities and cultural partners and also at a cross-sector international level of strategic advocacy. In this advocacy work, we share practice on what is effective to activate people through training and through the movements we support. Primarily, we work with Culture Declares Emergency (CDE), a global movement based in the UK, helping individuals and organisations to declare a planetary emergency and form local hubs to tell truths, seek justice and take action. Like CMUK, it is a distributed movement organised around principles of localism. Its regional and international hubs are communities of practice established by declarers collaborating in places to share knowledge and engage the public. This movement was co-founded and co-ordinated by the authors, with others. We in CMUK have worked with CDE to promote our Culture Takes Action framework, offering eight pathways for active response by cultural organisations. This aims to expand perceptions of the role of Culture beyond decarbonisation, which is an essential path but should be combined with decolonisation, adaptation, local transition, global sustainable development, cultural therapy, truth-telling and ecological innovation (Culture Takes Action Framework 2021). These pathways are framed within the challenge of Kate Raworth's Doughnut Economics, which is to create a safe, just space for humanity within the (mostly breached) limits of the Earth's operating systems (Raworth 2017).

It is unusual for a museum today to have no response to some aspects of the Earth crisis, particularly climate change, so there are countless examples of sustainability plans and education programmes. However, there are very few examples of activational museums that go beyond 'sustainable' to 'regenerative' or show awareness of collapse. A 'sustainable' approach is indicated by reducing or offsetting the harms of existing activities to sustain the organisation and its relationships. In business sectors, it is characterised by incremental actions that work more to save reputations than planetary systems. A 'regenerative' approach is indicated by dedicated missions and ambitious interventions that optimally contribute to regenerating social capacities, bioregional ecosystems and Earth's operating systems. This should be combined with being collapse-aware, which means anticipating and responding to the unfolding crises of ecological breakdown, pandemics, inequalities, displacement, famines and conflict.

Members of sector communities are starting to tackle the challenge of engaging communities with these problematic issues. Some, for example, in the Museums for Future network, are pledging to offer space to young activists and platform voices of affected people. The new manifesto of the Happy Museum Project (Jennings n.d.) calls upon museums to be agents of change in response to these multiple crises. The Museums Association has launched a campaign and resource hub on Climate Justice (Museums Association n.d.). There have been a growing number of conferences, journal issues and research initiatives on questions of curating and collecting climate, such as Reimagining Museums for Climate Action (Museums for Climate Action).

In all of these, we play a challenging role, asking questions such as: What does it look like, in this time, to be an activist and collapse-aware museum? What are all the different and emerging perceptions of what this means and the practices that arise from these perceptions? If the practice includes declaring a planetary emergency, does this trigger a shift in core purpose to *activating* people in places for transformative change, as we believe it should? Would we be more effective if we see people as inhabitants, nested within their bioregions and the biosphere, alongside other species?

What follows describes the practices we are experimenting with as associates and partners of CMUK.

Designing for transformative activation

In CMUK, we are designing experiences in diverse, experimental ways for 'affective participation', which includes using collections as transitional objects to stimulate safe but powerful conversations, inspire the imagining of alternative futures and support people to take action that prefigures the system changes needed.

Our core Theory of Change about activation is based on our hypothesis that transformative change happens when there is a critical mass of people thinking and acting for the benefit of places and planet, and benefitting in turn, working together over a sustained period in intimate relationships or geographical proximity. Leaders and decision-making systems are more capable of giving creative responses to ongoing crises to avert continued injustice and enable thriving diverse communities. The transformation needed in society is to help people equitably adapt to 'baked-in' harmful impacts of the planetary emergency and to be motivated to change systems to mitigate impacts still to unfold. We propose that activational interventions by museums should not be measured in terms of potential attitudes indicating behaviour change by individuals but rather in terms of increased capacities evidenced when people act together as caring planetary citizens. These capacities would be evidenced in social narratives, behaviour patterns and emerging policies and economic trends that are more ecocentric and equitable. To give an example, rather than an exhibition's impact being tested by the numbers of visitors who intend to reduce meat consumption, one might look for measures such as how many retail outlets nearby start offering more plant-based products. This approach raises the question of how museums can detect their contribution to changing social patterns alongside other interventions such as broadcast media or charity campaigns. The implication of evaluating impacts in terms of social patterns is that one must design interventions for such an impact and for the essential contribution of a museum compared to other forms of media, culture or community service.

Climate Museum UK is *experimental*, intending to demonstrate and advocate for what is possible in the Cultural sector, particularly focusing on museum practices, albeit taking a broad approach to the definition of museums. So this chapter aims to reflect on some existing practices that prefigure what more is needed and to illuminate the gaps by speculating where museum practice should be going. Extending from our core Theory of Change described earlier, the most effective activations by museums should therefore be those in which people are deliberately engaged to increase their agency, with consideration of tactics for reparative decolonisation, bioregional restoration and adaptation, and activism through planetary citizenship.

Partly our experimental nature is demonstrated by our *distributed structure* as a collective of creative practitioners activating our communities across the UK, helping people to express 'Earth emotions', explore the toxic histories of biocolonialism that led to the crisis and open their imaginations to possible futures. This engagement with local concerns is driven by the belief that people are more likely to understand the crisis as it relates to their own lives and to feel agentive in their communities. Our practitioners are collecting materials to engage local people as urgently as possible (rather than to collect for posterity) and, through this, also collecting people's responses.

The unique principle of CMUK is to be *possitopian* in thinking about the future. The possitopian approach expands the cone of possible futures and draws on geophysical realities and data while also harnessing the full powers of the imagination. Being possitopian means anticipating the future much more frequently, in many more situations and permutations, involving a greater diversity of people. Managing the risks of the planetary emergency is not about working out the best response to the most likely outcome and giving people simple takeaways to carry out in response. It is about determining the better range of responses to the full distribution of possible outcomes and empowering people to discuss and prefigure some of these pathways and to experiment with putting them into effect. In working with people in museological ways, we help them imagine future scenarios which are potentially more catastrophic or more abundant than they might allow themselves to think, drawing on experiences or evidence from the past and weaving these together in reference to real contexts. For example in a workshop we facilitated as part of We Breathe Together, a day on air pollution at the Horniman Museum in London, we used a selection of our handling objects to expand thinking on how air quality could be worsened not just with transport increases but by climate impacts and animal agriculture and then invited participants to visualise the safe, clean and green future they would work for by making collages on a paper forest displayed in the museum's conservatory.

While each practitioner works with local or specialised communities in the above ways, we are also a *digital museum*. Developing a digital presence allows us to collect responsively, inviting people to share stories and resources. For example, we invite and support articles, discussion and creative responses about Extreme Weather Stories, and about Everyday Ecocide.

As a collective, we have imagined a future state where Climate Museum UK has: *a culture* of sharing and living our six principles,[1] despite being distributed across several places; *capacities* between us to creatively engage our local communities with climate and ecology, expand imaginations and activate people to create regenerative places and increased *consciousness* among Cultural professionals to be more collapse-aware and pluricentric (Pascual et al. 2022).

Transformation starts with ourselves

There is a difference between experiencing concern about the Earth crisis and facing it. Facing it requires transforming one's psyche and relationships, bringing new awareness into active citizenship in your social, professional and community spheres. Coaching Scholar and Researcher Angélique Du Toit (2007) suggests that coaching as a transformative learning process is constructivist because clients are encouraged to continuously interpret their sense of existence to create meaning or develop alternative views. This is very familiar for those of us trained in constructivist museum education since the 1980s. Still, we must apply this now to a much less certain context where external changes are traumatic. It has been easy to construct learning around notions of human progress made possible by a stable biosphere but much harder to scaffold learning when the edifice is collapsing. We suggest that museum pedagogy has something to gain from critical coaching psychology. Critical coaching takes a non-directive approach; it builds a collaborative and facilitative relationship supporting self-responsibility and empowerment. It has a forward focus and is oriented towards action. While offering containment, it inspires critical thinking and reflective learning to help participants think and act differently (Shoukry 2016; Western 2012).

CMUK activations take a transformative learning approach (Mezirow 1997) that engages participants with their feelings about this crisis, supports critical thinking and facilitates locating where their agency lies. This is delivered through CMUK's 'feeling, thinking, doing' structure, an iterative triadic process designed to shift people from non-constructive coping strategies, to inspire visions of planet-kind systems and encourage action to construct this world they have prefigured (*Climate Museum UK* 2021).

Our practitioners use art, objects, ideas, games, books and natural sites to activate people to play, make, think and talk about the Earth crisis and open their imaginations. In all of our activities to date, since our first prototyping workshop in July 2018, we have found that the more open-ended and creative we have been in using objects, the more effective we are at opening up conversation and helping people find coping strategies and solutions, compared to when constraints have forced us to use more formal or instructional methods. For example, when the COVID-19 pandemic required us to use Zoom for our Creative Climate Conversations for Lewisham communities, we held up objects to the webcam and tried to elicit responses to them, but the conversations became diverted by more general questions or personal stories that reflected experiences that other participants could not relate to. However, during an in-person version of these workshops, objects could be handled, offered to each other as gifts and combined to enrich each other's storytelling. This enabled a sympathetic and open dialogue about the values that led to materials being extracted and wasted. Multiple playful combinations of objects have created a value that 'everything matters'. That is, each person's interpretation is carefully considered by the group and is just as valid as an opposing interpretation.

We take a psychodynamic approach to provide safe spaces for people to explore difficult topics, recognising that defence responses are comfort-seeking distractions from effective action, echoing the basic response of 'fight, flight or freeze' and that these can be socially constructed habits. CMUK practices are informed by object-relations' psychology, around which there is a growing body of evidence demonstrating the personal and social health benefits of haptic activity in museums, for example through handling and creative response to collections (Chatterjee et al. 2009; Froggett and Trustram 2014; Silverman 2009). However, unlike the intrapersonal and human societal definitions of well-being found in these papers, CMUK has an *ecocentric* understanding of well-being that includes the interdependencies of all species in nested ecosystems.

We are also informed by eco-psychology, such as the work of Dr Renée Lertzman. She has developed a model called the Golden Trine of Behaviour Change. It articulates the three aspects of people's responses to conflicting desires when faced with facts that significantly contradict their expectations for life and require them to change their behaviours and plans. The response of *being in denial* is about refusing to explore truths or being misinformed and then not being able to internalise and make authentic meaning from these truths. This stops people from even considering effective action. The response of *disavowal* may mean that people who have access to the facts are failing to make sense of the situation or to process it emotionally, so they are less motivated to take action. The *rationalising* response is often more accepting of science. However, by focusing so much on a linear, logical path, people assume solutions can be successfully applied without navigating social and emotional barriers, or they apply a single solution in their own lives, reasoning that it will be effective if everyone can follow suit. These responses mean that most people do not progress through a range of difficult emotions, do not embrace the complexity of diverse views or wicked problems and do not forge pathways that allow them to prefigure and create new ways of living.

Museums can support people through more constructive journeys after facing the Earth crisis due to their unique contributions in spanning human and more-than-human interactions over eras and geographies. Our profession needs to more finely co-design engagement programmes that can master this view over time and place. In light of Lertzman's Golden Trine, these programmes can look in three time-based directions: the past, to learn and discover truths and histories; the present, to make sense of emotional and social aspects, supporting mutual care and resilience during crises and the future, growing an appetite for regenerative system change.

Activation through feeling, thinking and doing

As a collective of practitioners, we frequently reflect on the activities we deliver and apply this learning to new projects and workshops. As already noted, our optimal model takes participants on a 'Feeling, Thinking, Doing' journey, which is cyclical by starting and ending with self-reflection. The process is: embodied sensing around objects or images, or a site; expressing feelings and listening to others; thinking about them in a wider context of histories or future scenarios; practising action by mending, making, playing or planning; returning to the self to make pledges.

Next, we break down the three main components with examples of practice.

Feeling

We create 'holding spaces' or 'potential space' for participants, who might be either public or professional audiences, to explore complex and challenging aspects of the Earth crisis through interpersonal communication, creativity and play. We are influenced by Winnicott's articulation of play being a third kind of reality that is neither the inner reality of the participant nor the outer reality they engage with but the potential in the act of creating (Winnicott 2005).

We often use museum collections (or our own artworks or found items) as transitional phenomena, where participants are invited to interpret objects and make their own. They express personal meanings when they connect objects with material experiences in their own lives or their questions about the world. For example, in our Wild Museum pop-ups, animal curators show their extracted collections of ceramics, metal, coal and imported seeds, then invite participants to make a wilder museum of clay objects. In using hands to manipulate a primordial material, participants are able to speak freely. In other workshops, we might invite participants to make an object to represent their conception of the Earth crisis, or to document their bodily senses in response to a river, for example (Figure 3.2).

While the opening parts of our designed experiences focus on emotions and attitudes, we might use key facts or science-informed scenarios as stimuli for those expressions before moving on to deeper discussion. Facing truths – whether

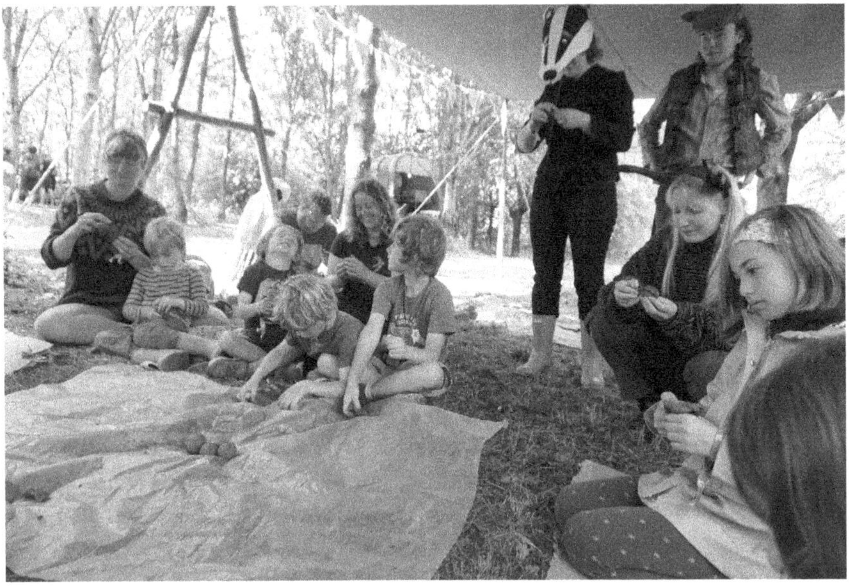

FIGURE 3.2 The Wild Museum popping up at the Timber Festival, July 2021

through metaphors or firm science – is a first stage towards activation. We believe museums should respond to growing public environmental concern by creating space for nuanced conversation around these truths, exploring biases, narratives and conundrums.

Thinking

We hope to guide participants through a journey of discovery and revelation by: providing access to current science; exploring the lived experience of climate impacts; investigating big systems of geophysics and human power; and delving into multiple perspectives on solutions, concepts, ethics and terms.

One of our enquiries as a collective of practitioners is how to expand audience perspectives to see interconnected emergencies and their causes. This enquiry originated in our Stories of Extraction project in McKenzie (2022) when we were in the first cohort of Activist Museum Awardees. Together, we are considering how to make the following more visible: the entwining of colonialism and environmental extractivism over centuries; interconnections between direct ecocide and climate breakdown and how toxicity arising from the extractive economic system leads to poor health and immunity via polluted air, highly processed food, sedentary lifestyles and zoonotic diseases.

We do this in various ways, for example, by working with partner museums and their communities to re-story their collections. Our training programme 'An

Eco-lens on Things' works with museum professionals to creatively explore the significance of their collections from ecological and trauma-informed perspectives. In the words of coaching psychologist Simon Western (2012), we 'look awry' at them by adopting alternative eco-lenses, such as the viewpoints of other animal species, including microbiota; from the long-term stance of a person in the future or from the perspective of water. Participants are invited to critically reflect on the multiple stories behind their collection objects or sites, using the lenses to shift from familiar mindsets to rethink and expand narratives of humanity in relation to nature. For example we might ask, what would happen if bees were to create a story from a museum collection?

From this training, many museums have begun inviting people to see their collections in a new light. From a mentoring session with Royal Cornwall Museums, a case of objects was installed in the high street to spark creative climate conversations. After a training session with staff from museums across India, the participants formed Indian Museums Against Climate Change and are exploring many ways to connect their diverse collections to explore topics such as food and fashion.

Tullie House Museum and Art Gallery in Carlisle has been in partnership with us on their Once Upon a Planet programme (Once Upon a Planet 2022). We delivered training workshops for their whole team, putting eco-lenses on their natural history, geology and local history collections. We recruited Megan Bowyer, a young artist practitioner, to 'look awry' at their collections, to run creative workshops and to create artworks to incorporate into their temporary exhibition, 'Human change not climate change'. As part of her residency, Megan helped staff and audiences think about how we subjectively interpret nature by creating artworks from an archive of natural history observations by local volunteers. She created artworks that combined items from across museum categories, such as bird egg shells and doll's house China, stirring thoughts about fragility, durability and natural materials in industrial processes.

Doing

In our triad's 'doing' element, participants are guided to move their learning into transformative action. This is the most challenging aspect of our activities because time is limited, and participants become absorbed in articulations of complex issues. We always try to build in an active phase, whether this is drawing a solution, planning next steps with colleagues or creating a future vision.

Optimally, we aim to support participants to locate where they can be more planet-kind in their domestic, social, professional and civic spheres and exert change on harmful systems. All our associates and partners have access to our People Take Action framework (Figure 3.3), which provides scaffolding for people to identify and expand their areas of agency.

It offers eight pathways for people to act in any area of influence they have, expanding their notions of what action is possible. It sees all actions as 'personal

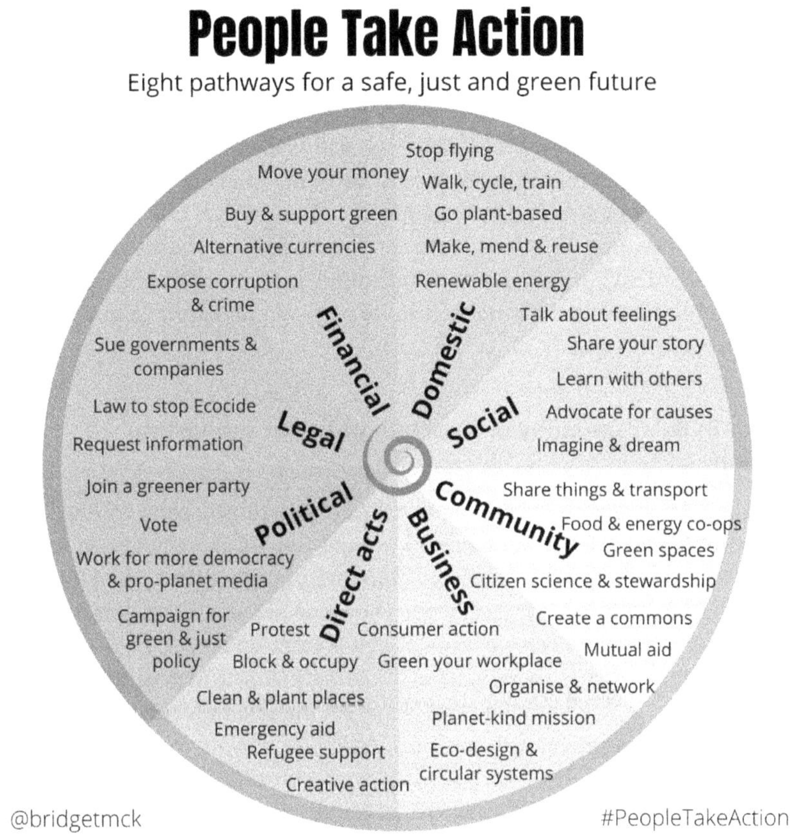

FIGURE 3.3 People Take Action framework used in workshops with public audiences

actions' within our blended lives as consumers, friends, citizens and workers. We are piloting workshops using this framework.

It is commendable when museums make efforts to activate visitors after experiencing environmental exhibitions. For example the Museums of the University of St Andrews created the Dive In! exhibition, working with Adam Corner, climate communication expert (Dive-In Exhibition 2022). They devised four personas, characterised by a marine animal that visitors could identify with, ranging from eco-curious to eco-activist. While the personas were less effective, as visitors preferred not to pursue a route with one identity, many were willing to follow up with the suggested actions that were most clearly defined. Similarly, Wakefield Museum offers a pledge tool shared on social media, showing people simple actions they can take under several headings. However, there is a risk with simple calls to action that, although there are more measurable results, impacts may be limited to domestic spheres of action. It may also reinforce a 'single action bias' whereby each of us has a 'finite pool of worry' ameliorated by repeating a beneficial but easy action as a replacement for more agentic organising (CRED Guide: The Psychology of Climate Change Communication 2022).

One way museums could tackle these limitations is to model a larger-scale action that makes the best use of their assets and is locally appropriate. From this grand action, they can inspire audiences to carry out many other actions that will transform their communities. For example, the Woodbridge Tidemill is setting up a microgeneration project, using water to power devices and as an educational resource (Ballard 2022). This could, for example, inspire community solar co-ops, hydropower design competitions, the revival of local grain milling or schools running sustainable bakeries.

'A Space To Be': case study of a partner museum supporting activation

Scarborough Museums and Galleries have launched a project called A Space To Be in a room of their Art Gallery, dedicated to exploring local regeneration and well-being. After a year-long pilot, processes from this project will be embedded into the organisation's evolution in the long term. The value of a long-term approach is that change in the community can emerge through bottom-up processes.

Its overarching aim is to support active citizenship in regenerative practices and communities in Scarborough. This sits in the context of North Yorkshire's aim to become carbon-negative, with business initiatives moving towards regenerative economies (Pidd 2022). The Space To Be was conceived to support the community post-COVID when there was a sense that social regeneration was inseparable from environmental regeneration. The project's ethics were influenced by CMUK's Culture Takes Action framework, broadening the understanding of well-being to the 12 social foundations of Doughnut Economics, which must sit within ecological limits (Raworth 2017).

A Space To Be is multifunctional and designed for different groups to feel comfortable to be creative and share stories about their place. The tenets for using this room, based on reciprocity, are co-created with participants who drew up a 'manifesto' for its use. Groups booking the room sign up to this 'manifesto' and answer questions such as: how does your use of this space contribute to its principles, and how do they enrich your group?

Reciprocity principles also inform the materials and activities. User groups are invited to respond to provocations stimulated by gallery collections about regenerative culture, which are offered in a co-created zine and animated story. Participants are encouraged to leave gifts and art materials, creating an enriched environment for future visitors.

Projects and activities in A Space To Be are designed with reference to CMUK's iterative process of 'feeling, thinking, doing'. They are facilitated in a non-prescriptive way to inspire reflection on nested layers of well-being, which extend widely to environmental systems while starting from participants' personal experiences and views.

Projects include 'Map it Well', a mindfulness map of Scarborough with sensory and imaginative journeys created from community knowledge. Routes on the map might follow a colour or sound or the perspective of an animal interacting with the environment. Future iterations will dig deeper into the interdependencies of well-being. 'The Hello Quilt' is a collaborative quilt-making project, welcoming drop-in visitors, including refugee groups and those struggling to heat homes.

Another project will tie in the idea of 'solastalgia' with the natural history collection, exploring eco-anxiety and nostalgia for lost environments. It will bring together older generations with memories of thriving nature and younger people who fear future loss and damage. Recent threats to legal protections of nature and the joined-up response of conservation charities can motivate conversations and broaden engagement in activism by the museum's communities.

By progressing an understanding of well-being that considers community and environmental interdependencies and by modelling reciprocity as a vehicle, the project's impacts should spiral beyond the walls of the Space, creating active citizenship for other regenerative initiatives in the region.

Conclusion

Museums are responsible for retelling stories and facilitating the visioning and birthing of alternative futures. Stories consist of a dynamic understanding of past, present and future. If we can critically reflect on the past, we see that current systems are not inevitable or fixed. Andrew Simms, in his article 'Museum of Rapid Transition: museums in a world facing existential crisis', proposes *"an inventory of how societies have achieved rapid transitions in the past may begin to codify for us the ingredients, or broad design criteria, for successful future rapid transitions"* (Museum-ID n.d.).

We support this role of museums to be transformative, articulating multiple stories and creating inventories to inform and inspire collective action in places, which might include running mutual aid groups; co-operative supply of renewable energy; promoting regenerative agriculture; advice on retrofitting heritage buildings; offering nature-connected education; or promoting car-free streets through education about urban heritage in the past.

For museums, the ubiquitous and uncertain transformations to planetary systems challenge the notion of continuity behind their founding rationales. Everything has changed. Planet Earth is becoming a museum in itself, displaying the traces of systems, cultures and species coming to an end. Museums can hold a mirror to this global museum of endlings, highlighting the harmful systems as human-constructed rather than as an inevitable feature of human progress. The challenges for museums include reflecting this planet-scale change within their walls, overcoming political resistance to radical thinking, and creating satisfying and accessible narratives for our audiences. In CMUK, our lack of a building and valuable objects encourages us to look outwards and distribute our focus across many objects, people and their places. Given the tentacular nature of the emergency, we model our own proposition of abandoning the grand narrative, instead adopting a pluricentric approach whereby every object or site in every interaction with people can open up fractal possibilities to understand the past and imagine different futures.

Note

1 Our six principles are Compassionate, Holistic, Intersectional, Participatory, Possitopian and Planet-kind.

References

Ballard, S. 2022. *We Install a Micro Generator! Woodbridge Tide Mill*. https://woodbridgetidemill.org.uk/2022/08/15/we-install-a-micro-generator/ (Accessed 29 September 2022).

Cassely, J.-L. and Fourquet, J. 2020. *La France: Patrie de la Collapsologie*. https://www.jean-jaures.org/publication/la-france-patrie-de-la-collapsologie/ (Accessed 8 August 2023).

Chatterjee, H., Vreeland, S. and Noble, G. 2009. Museopathy: Exploring the Healing Potential of Handling Museum Objects. *Museum and Society*, 7: 164–177.

CRED Guide. n.d. *The Psychology of Climate Change Communication*. http://guide.cred.columbia.edu/guide/sec4.html (Accessed 29 September 2022).

Culture Takes Action Framework. 2021. *In Climate Museum UK by Bridget McKenzie*. https://bridgetmck.medium.com/culture-takes-action-framework-29e8dfaaba24 (Accessed 28 September 2022).

Dive-In Exhibition. n.d. *MASTS*. https://masts.ac.uk/cop-26/dive-in-exhibition/ (Accessed 29 September 2022).

Du Toit, A. 2007. Making Sense Through Coaching. *Journal of Management Development*, 26: 282–291. https://doi.org/10.1108/02621710710732164

Froggett, L. and Trustram, M. 2014. Object Relations in the Museum: A Psychosocial Perspective. *Museum Management and Curatorship*, 29: 482–497. https://doi.org/10.1080/09647775.2014.957481

Galey, P. n.d. *World May Breach 1.5C Warming Within 5 Years*. https://phys.org/news/2021-05-world-breach-15c-years-wmo.html (Accessed 24 September 2022).

Global Assessment Report on Disaster Risk Reduction. 2022. https://www.undrr.org/publication/global-assessment-report-disaster-risk-reduction-2022 (Accessed 24 September 2022).

Jennings, H. n.d. *Share Your Reflections on Our New Manifesto – Happy Museum Project*. https://happymuseumproject.org/our-happy-museum-manifesto/ (Accessed 24 September 2022).

Lertzman, R. 2009. *The Myth of Apathy: Psychosocial Dimensions of Environmental Degradation* (PhD). Cardiff: Cardiff University.

McKenzie, B. 2022. *About Stories of Extraction: Stories of Extraction*. https://medium.com/stories-of-extraction/about-stories-of-extraction-45d3ab8d6d98 (Accessed 24 September 2022).

Mezirow, J., 1997. *Transformative Learning: Theory to Practice. New Directions for Adult and Continuing Education 1997, 5–12*. https://doi.org/10.1002/ace.7401

Museums Association. n.d. *Museums for Climate Justice – Campaigns*. https://www.museumsassociation.org/campaigns/museums-for-climate-justice/ (Accessed 24 September 2022).
Museum-iD. n.d. *Museum of Rapid Transition: Museums in a World Facing Existential Crisis*. https://museum-id.com/museum-of-rapid-transition-the-role-of-museums-in-a-world-facing-existential-crisis/ (Accessed 8 August 2023).
Museums for Climate Action. https://www.museumsforclimateaction.org/ (Accessed 24 September 2022).
Once Upon a Planet. 2022. *Tullie House Museum & Art Gallery*. https://www.tulliehouse.co.uk/events/once-upon-planet (Accessed 29 September 2022).
Oxfam International. 2021. *Six-Fold Increase in People Suffering Famine-Like Conditions since Pandemic Began*. https://www.oxfam.org/en/press-releases/six-fold-increase-people-suffering-famine-conditions-pandemic-began (Accessed 24 September 2022).
Pascual, U., et al. 2022. *Summary for Policymakers of the Methodological Assessment of the Diverse Values and Valuation of Nature of the Intergovernmental Science-Policy Platform on Biodiversity and Ecosystem Services (IPBES)*. https://doi.org/10.5281/zenodo.6522392
Pidd, H. 2022. North Yorkshire Puts Seaweed at the Heart of Its Carbon-Negative Ambitions. *The Guardian*, 27 September 2022.
Pörtner, H-O., et al. 2021. *Scientific Outcome of the IPBES-IPCC Co-Sponsored Workshop on Biodiversity and Climate Change; IPBES Secretariat, Bonn, Germany*. https://doi.org/10.5281/zenodo.4659158
Raworth, K. 2017. Why It's Time for Doughnut Economics. *IPPR Progressive Review*, 24: 216–222. https://doi.org/10.1111/newe.12058
Shoukry, H. 2016. Coaching for Emancipation: A Framework for Coaching in Oppressive Environments. *International Journal of Evidence Based Coaching and Mentoring*, 14: 15–30.
Silverman, L.H. 2009. *The Social Work of Museums*. London: Routledge.
Steffen, A. 2021. *Planetary Crisis as Transapocalypse*. https://medium.com/@AlexSteffen/planetary-crisis-as-transapocalypse-3db86d512c49 (Accessed 24 September 2022).
Stokes, C.R., et al. 2022. Response of the East Antarctic Ice Sheet to Past and Future Climate Change. *Nature*, 608: 275–286. https://doi.org/10.1038/s41586-022-04946-0
Warszawski, L., et al. 2021. All Options, Not Silver Bullets, Needed to Limit Global Warming to 1.5°C: A Scenario Appraisal. *Environmental Research Letters*, 16: 064037. https://doi.org/10.1088/1748-9326/abfeec
Watts, J. 2020. James Lovelock: 'The Biosphere and I Are Both in the Last 1% of Our Lives'. *The Observer*, 18 July 2020.
Western, S. 2012. *Coaching and Mentoring: A Critical Text*. Thousand Oaks, CA: SAGE.
Winnicott, D.W. 2005. *Playing and Reality*. London: Routledge.

4

MUSEUMS AS CATALYSTS OF CULTURAL ADAPTATION

The 'Inside-Outside Model'[1]

Douglas Worts

Introduction

In theory, it is within humanity's ability to build a global culture in which more than eight billion people can be healthy, happy, compassionate, creative and adaptive on Earth (Worts 2011). People coming from countless backgrounds and ancestries have the potential to join together in the spirit of equity and well-being for all. They can share insights from the past in anticipation of making meaningful, conscious and timely decisions in the present that aim to foster well-being of both people and planet. It is even possible for societies to envision new ways of living that support the Earth's ability to continuously regenerate itself (Wahl 2016). In order to turn theory into practice, humans need a fresh way to think about our living culture and how it needs to change, especially in this time of ecological, social and economic crisis (Homer-Dixon 2006).

Early in the 1980s, the United Nations created the World Commission on Environment and Development (aka the Brundtland Commission). Its 1987 publication, *Our Common Future*, introduced the concept of 'sustainable development' as a way to 'meet the needs of the present without compromising the ability of future generations to meet their own needs' (Brundtland 1987). Since then, the term 'sustainability' has been used to rationalise all sorts of counter-intuitive behaviours, including the *sustaining* of the status quo. The obfuscation related to the multiple interpretations of this term has disoriented business, government, economics, education and much more, including the museum community.

For more than eight billion people to live and thrive on this planet, it is essential that each generation is able to 'meet their own needs', while ensuring that future generations can meet theirs. Sadly, humanity has created no consensus vision of such a future. Instead, there is often an unarticulated assumption (at least among

DOI: 10.4324/9781003347606-5

the dominant players holding power) that the status quo cultures that have created the world we currently occupy are fundamentally working well. There is some acknowledgement that 'business as usual' has in the past, and does in the present, created unintended impacts that have caused significant environmental and social damage (e.g. social and economic inequity; a climate crisis; ecosystem damage; ocean health in decline; unprecedented species loss and much more). If humans can muster the courage to step back and examine this situation, we may learn that our culture(s) need to transform themselves in ways that enable our species to exist in a balanced relationship to local and planetary systems (Sutter and Worts 2005; Worts 2003). We also may learn that a pre-condition for such balance is ensuring equity for all, across our species.

Historically, humans have been good at adapting to an ever-changing world. Disruptions from extreme weather, earthquakes, forest fires, droughts and the like have all prompted humans to act in new ways. Such adaptive behaviour enabled them a chance to survive and even thrive. Since humans first evolved, they learned to leverage materials and processes in nature to enhance their quality of life. Energy was harnessed from wood, animals, wind, rivers and so on to accomplish astounding tasks. Technology was developed to extend the reach of humans into many worlds that had been unfathomable – from exploring the microscopic to charting the limits of the universe. The atom was not only identified but also eventually split, unleashing unimaginable power that has been used for both good and evil. The creativity of human beings has made our species unique in the story of life on this planet. But it is essential that, at this moment in human history, we revisit what notions of 'success' we use to guide our cultures (AtKisson, 2010). The lived experiences of the past must be used to generate insights that can help us understand our present realities and to imagine a viable and desirable world that will serve our species in ways that enable humans to continue living within the life systems of Earth. While our past may have been energised by competitive approaches that promised personal reward, in terms of money and power, our vision for the future must revolve around the well-being of humans, embedded within the rich web of life that makes up our world – and which is not only threatened but arguably already in decline – because of the unconscious and dysfunctional culture of our contemporary world. Humans are systematically, if unintentionally, disrupting the Earth's biodiversity and natural systems – including climate and weather, ocean chemistry and soil health.

As museums increasingly muse about what it means to be 'cultural organizations' in the 21st century, our profession has the ability to reassess its traditional practices and develop new ways of serving the imperative of human adaptation in our extremely precarious world. In fact, adaptation is no longer a sufficient goal. Unless we can find ways to actually regenerate the natural capital that has been lost at the hands of 'progress', we will continue to witness the decline of the environment's resilience. Now is the time for museums to confront one of the greatest challenges they have ever encountered – specifically bridging the gap between the living culture and institutionalised culture. This gap is enormous – however, it may not be as hard to address as it first seems.

The quest for striking a dynamic balance between humanity and the complex natural systems of Earth has never been part of traditional museology. Even today, few museums assess and measure their impacts on the living culture. Some may conduct a certain amount of audience research, but this is largely focused on assessing visitor satisfaction of on-site experiences in the leisure-time economy. Very few museums assess the ever-changing needs of the increasingly urban, pluralistic and globalised cultures of our world – let alone assess whether institutional programming is actually meeting those needs. However, if museums use experimental approaches which plan for impacts and outcomes at a range of cultural levels (e.g. individuals, communities, cities, regions, societal systems), rather than outputs for individuals and families to consume within the leisure-time economy, then a new era of museums may be upon us (Worts 2016, 2017).

The purpose of this chapter is to introduce a museum planning tool, called 'The Inside-Outside Model: Museums Planning for Cultural Impacts' (I-O Model), which was designed to help museum professionals to work co-creatively, across the culture, to envision and realise futures that are both scientifically viable and ethically desirable. It should be understood that this approach is somewhat heretical. There is little focus on collections, exhibits, on-site programs or engagement with tourist markets – although all of these will surely be part of what museums do as they work with the public as catalysts to generate meaningful cultural impacts.

It may be useful to remember that today's sustainability challenges are not simply corporate problems of operational efficiency that can be addressed by making existing operations 'less bad'. While being 'less bad' is always better than being 'more bad', this is not the nature of fostering a living culture of sustainability. Instead, museums will benefit from reflecting deeply on the many ways in which they can foster adaptive approaches across the living culture, not simply on the museum site itself. Before introducing the I-O Model, it is worth identifying a few of the unique challenges and opportunities associated with our moment in history.

The challenges of our time

Never before has a single species pushed the Earth beyond its ability to regenerate itself. Never before has a single species dominated, and often damaged, so many other species and their habitats. However, on the other hand, never before have we seen the kind of creativity and problem-solving in any species, other than humans. And yet, the strategies that humans have developed to realise their visions, and to solve big challenges, have ultimately failed us. Systems of governance, economics, technology, religion and more have largely proven to be inadequate over time, especially when scaled to global levels. The living culture is multi-levelled, timely, archetypal, contradictory, affirming, messy, creative, always changing, partly conscious and partly unconscious. In many ways, living culture is the opposite of the tidy explanations that are so often the mainstay activities of traditional museums.

For many years, powers over how humanity has evolved were largely in the hands of governments, business, religions, as well as powerful individuals. The result has been massive growth in global population, inequality, migration, urbanisation, industrialisation, pluralisation, globalisation and more. Sadly, the growth of our species has not been guided by the necessary wisdom to ensure that human life remains within the parameters required by Earth's natural systems. Creating human systems that increasingly upset planetary balance is a perilous path (Janes 2009). In the past, cultures were often reasonably successful at assessing negative impacts on local ecosystems, which in turn enabled communities to adapt. However, in more recent times, we have witnessed the expansion and relocation of industrial production to parts of the world where business goals of 'economies of scale' production, reduced costs and fewer regulations all contributed to the lure of increased profits. At the heart of this phenomenon is an economic system that demands endless growth in resource consumption and the centralisation of wealth, which have ultimately presented us with existential threats to humanity's own well-being, as well as that of other species.

Human survival, and even thriving, remains possible. However, this potential demands the adaptation of current systems in order to create balance in the larger world. For humans to remain on our current path is to risk losing everything. The following is a list of some of the major trends that define our time, and which must be redirected towards safe harbour, if our future prospects are to improve.

The Anthropocene

Approximately 75 years ago, humanity entered a new geological period – informally known as the Anthropocene. The name and exact start date of this period are not yet finalised; however, a global team of geologists is currently considering these details. There seems to be agreement that the Anthropocene will be an 'epoch', which indicates that it is less than a 'period', but more than an 'Age'. The significance of this new epoch is that it is characterised by humanity having become the largest single factor in how planetary systems are changing (Koster 2020). The Anthropocene signals that the context for human life on Earth has taken a fundamental turn.

For the past 12,000 years or so, humans were able to exploit the wealth of Nature without causing more than local disruptions to natural systems. This relatively stable period is known as the Holocene, which followed the last Ice Age.[2] However, in the middle of the 20th century, exponential growth in human population, coupled with the ballooning scale of our resource consumption, and vast waste production have all meant that our species has become the number one force shaping Nature.

Global/local culture versus planetary boundaries

For a very long time, humans have been creative forces that have used the resources of nature to address their needs and wants. Humans have analysed situations and found ways to exploit available resources. There have always been unexpected impacts of this enterprising spirit, but until recently, our planet has had a massive capacity to regenerate itself and to reprocess pollution into useful materials. However, the sad truth is that the planet's regeneration ability is not limitless. From the mid-20th century onwards, humanity has been systematically violating the 'planetary boundaries'.[3] These boundaries involve large, dynamic systems that require relative balances to be maintained if there is to be overall planetary stability and health.

If these boundaries are not kept within prescribed limits, then planetary systems shift. For example climate change is one such boundary. The point here is that it is the largely unconscious behaviours and systems of cultures that are generating the activity that is violating planetary boundaries, while human feedback systems are failing to prompt adaptive changes. Given this dire situation, important questions need to be asked. Can museums transform themselves sufficiently to become catalysts of reflection, dialogue and co-creative action in the living culture? To what extent do the legal parameters of incorporated museums prevent the museum field from transforming itself so that it plays a more productive and urgently needed cultural role? What new roles could museums develop to improve the relationships humans have with both humanity and planetary systems? What are the opportunity costs of museums trying to address climate change primarily through operational efficiency measures, without prioritising and optimising their potential for generating meaningful impacts across the living culture?

Politics and business – at odds with adaptive cultural change

If we scan the world for examples of where political and business actions are creatively addressing our multiple planetary crises, there are few convincing examples. However, there are areas of inspiration that warrant examination. One is the work of Kate Raworth, a UK economist who developed the Doughnut Economics Model – which imagines replacing the traditional economic focus on continuous financial growth (Gross Domestic Product) with a commitment to using systems thinking to generate net-positive value generation across social, environmental and economic domains (Raworth 2017).[4] Raworth's revolutionary approach has also nurtured a global research and development think-tank, Doughnut Economics Action Lab (DEAL), which is conducting projects in many parts of the world to help clarify what it means to build a 'wellbeing economy'.[5] There are also many businesses that are committed to building enterprises that aim to generate social, environmental value within a viable economic operating framework (Klomp and Oosterwaal 2021).[6] And, inspiringly, New Zealand's recent Prime Minister, Jacinda Ardern, led

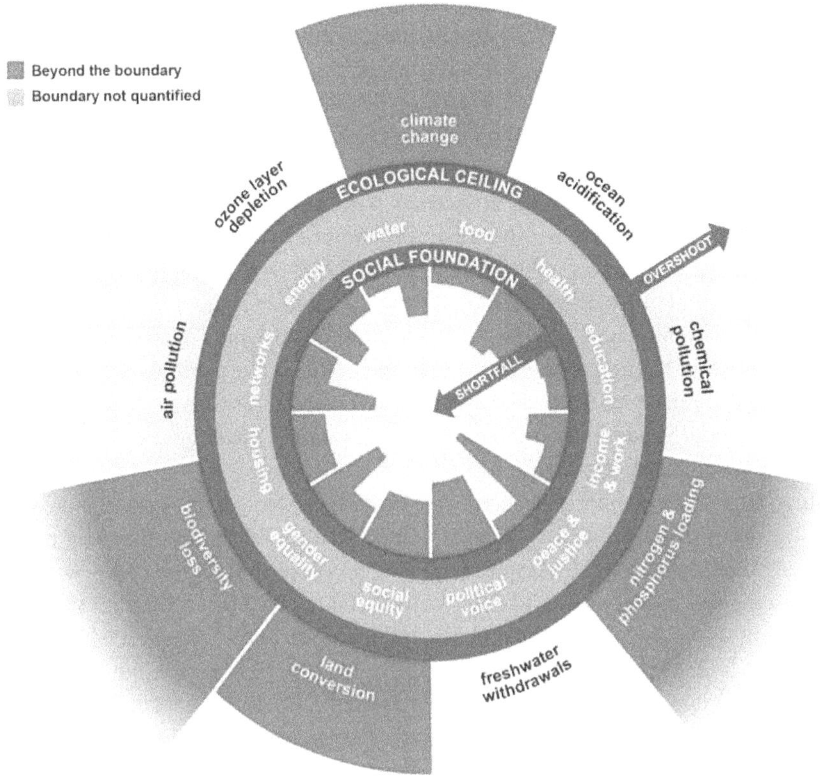

FIGURE 4.1 Doughnut Economics Model by Kate Raworth – economy that operates between a social foundation and an ecological ceiling. Photo credit: Wikimedia Commons – https://commons.wikimedia.org/wiki/File:Doughnut-transgressing.jpg

her government to declare that it would shift its national budgeting process away from a focus on GDP, towards a focus on environmental and human well-being.

The opportunities of our time

Humanity needs foundational cultural change to thrive, or even survive – but we don't have agreement on what such a culture looks like. Ecomuseums are oriented to moving communities towards well-being, however traditional museums are not (DROPS website). Imagining how museums could change in order to be effective catalysts of cultural change and adaptation – especially in the Anthropocene – is vital (Riva et al. 2022; Worts 2017). One aspect of how museums can catalyse change is through the co-creative partnerships that they forge.

Co-creativity is a powerful process that many museum professionals already understand well. Educators are perhaps most familiar with the process, because education is always co-creative whenever a teacher fosters in students the ability to 'make meaning' that draws on their own personal experiences, vision and associations. When there is a trusting bond between teacher and student, the latter's creativity is unleashed in new and often unexpected ways. It can result in new learning for both teacher and student. If a museum partners with a vision/values-aligned organization, and if there is a trusting, collaborative bond established, then the synergy can produce ideas, visions, insights and idea-generating tools intended to challenge current thinking patterns. In the event that such an approach was focused on the issues of our day (i.e. issues of the Anthropocene), then measurable impacts can conceivably be produced within the living culture. The significant point, however, is that if museums are to become catalysts of cultural change, their measures of success would need to be oriented to changes within the larger, living culture – not simply within museum buildings.

It is vital to remember that many museums have built great expertise in very specific areas of concern such as history and science. While expertise is a potent building block of human development, it may have come at a high cost – wisdom. While expertise uses narrow and deep focus to master the inner workings of things, wisdom involves the ability to step back and integrate knowledge and understanding from a wide range of experience. Expertise tends to be authoritative, while wisdom is more humble and open. Both are required – however, wisdom now seems to play second fiddle to expertise. The goal of expertise is control, while the goal of wisdom is well-being. Museums have the potential to facilitate the intersection of wisdom and expertise. Through such integration, museums can help cultures imagine flourishing, inclusive futures.

In 1972, insights and wisdom were offered up to humanity through the book *Limits to Growth*. In it, a group of scientists analysed population, consumption and environmental trends that anticipated the crises we see today, including climate change (Meadows et al. 1972, 2004). Their projected image of planetary systems' degeneration and collapse was about as sobering as one can imagine. And yet, even when presented with accurate insights into threats associated with 'business as usual' approaches, governments, economists and business leaders were dismissive of the warnings. In our current era of misinformation and conspiracy theories, we have unfortunately learned that science and facts are not enough for humans to act responsibly, courageously and with the interests of everyone in mind. When wisdom helps to marshal expert insights, and shape them into visions of viable and ethical futures, it is an essential process. When wisdom has no place, chaos soon emerges.

There are many ways to bring about systems' change – and if museums are to become catalysts of cultural adaptation, they will need to become very familiar with such processes, beyond their special expertise in traditional academic disciplines.

What may lie at the heart of 'culture', especially in the Anthropocene, is finding new ways to ensure that the well-being of the entire planet and all of its inhabitants remains the overarching vision of humanity. Figuring out how museums need to change in order to help realise such a vision will be a challenge – but what are the alternatives?

One of the central opportunities for any museum that intends to become a cultural catalyst is to expand its focus from generating cultural **outputs** for public consumption (exhibits, programs, publications) to facilitating processes of public engagement and co-creation that have meaningful **outcomes/impacts** on individuals, groups, communities, organizations and more. Needless to say, this amounts to a sea change in the vision and practice of museums in society. Accordingly, it will require the self-selecting few museums to begin working in new ways and then assess and report their impacts widely.

Luckily, there are already models of this approach within the museum world. Specifically, ecomuseums were conceived to serve the well-being of humans living within a region (Davis 2011; Riva 2017).

It is within this thought about museums becoming catalysts of adaptation in the living culture that the *Inside-Outside Model: Museums Planning for Cultural Impacts* (I-O Model) was created (see Figure 4.2).

I developed the model in 2019, initially to help guide conversations within the Sustainability Task Force of the American Association for State and Local History. Through its Task Force, the AASLH aimed to foster museum awareness, engagement and action related to sustainability (Worts 2019). The I-O Model was created to help manage two competing notions of sustainability. The first was 'sustainability' as a holistic balancing of multiple, interdependent, complex systems that currently are collapsing. The second was sustainability as 'greening' – which involves making the status quo 'less bad'.

In the following introduction to the I-O Model, there are two fundamental component parts. The first is the 'Inside' dimension, which focuses on the physical manifestation of the museum and its contents, as well as the governance, skills, knowledge, wisdom, processes and passion that are held by its staff (both paid and volunteer). The second dimension of the model is the 'Outside', which involves all of the component parts of our living culture – people, community, place, processes, values, goals, behaviours, systems, trends and more. Culture, in all of its forms and manifestations, lives throughout the 'outside' dimension.

The purpose of the model is to suggest ways that museums can leverage inside assets and processes, in order to support the complex, co-creative, cultural transformation needed to adapt in a changing world. With this goal in mind, the process is ever-evolving. It requires humility to understand that cultural adaptation can't be controlled as a top-down, mechanistic process. To ensure that people don't feel left out, it is best to design inclusive and supportive processes. Needless to say, this task is not easy.

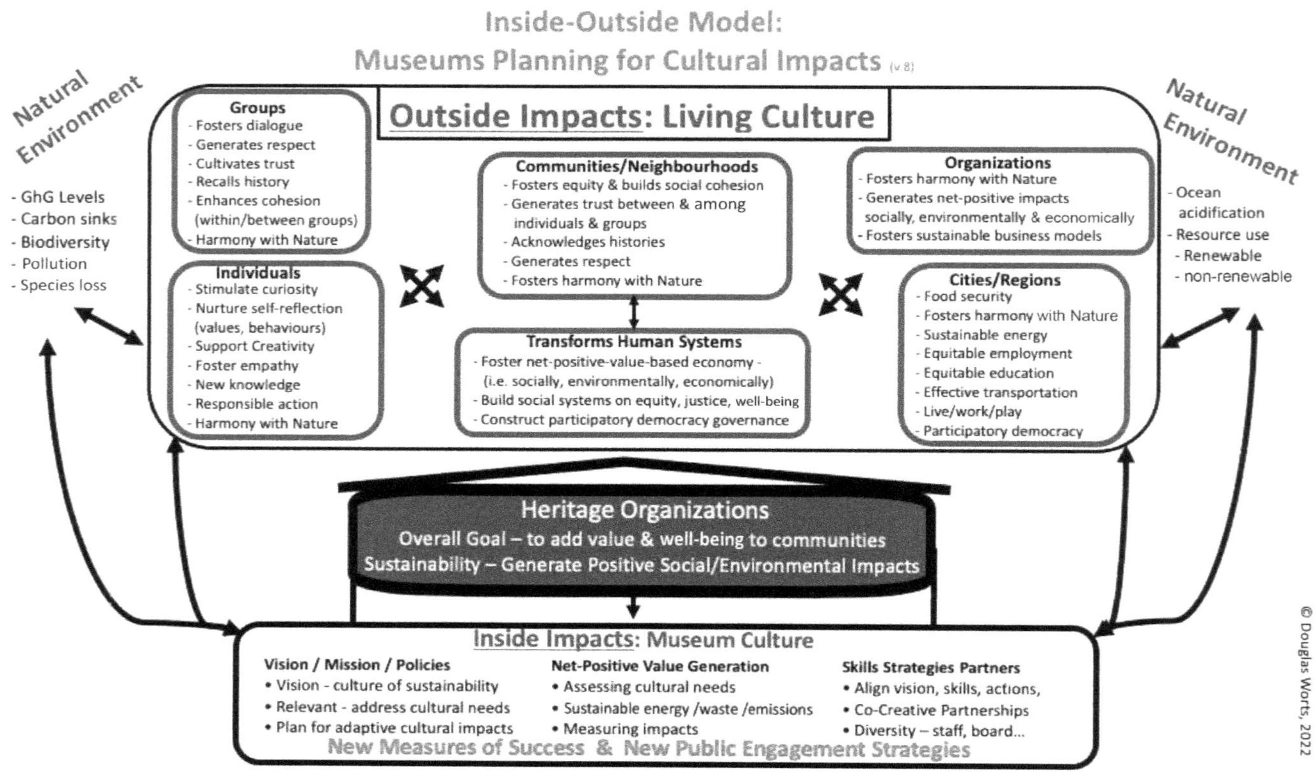

FIGURE 4.2 The Inside-Outside Model: Museums Planning for Cultural Impacts

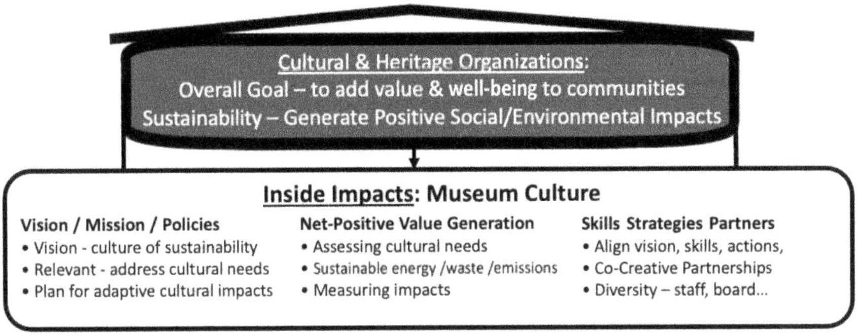

FIGURE 4.3 I-O Model – inside dimension (the museum)

The contents of the boxes are suggestive and designed to spur conversation and customization; they are not intended to be prescriptive or complete. Let's begin by examining the museum itself (see Figure 4.3).

In the most generic sense, public cultural organisations exist to serve the public good, in ways that add value and quality of life to their community. Surprisingly, museums are often vague about the ways that value is added and community well-being is re-enforced. Ideally, when cultural organisations aim to focus on sustainability, the impacts should be seen as adaptive change to both social and environmental aspects of the community.

The Inside dimension of museums is a highly organised, and often hierarchical, environment. Often adopting a corporate form (usually non-profit), it is normally guided by a vision and mission, as well as its stated values, traditional practices and policies. In addition, people with specific sets of skills are engaged to carry out what is normally considered core activities of these organisations. The privileged skill-sets, including discipline-based expertise related to collections; educational processes; public program development; partnerships; conservation of collections; organisation and management; marketing; needs and impact assessment and more, can all help to design the Inside dimension in ways that optimise desired impacts in the Outside Dimension (Hirzy 1992). If the goal is to foster an adaptive living culture that is aligned with the vision of a sustainable future, it will require astute mobilization of Inside resources and processes. Also necessary will be adept approaches to forging creative, vision/values-aligned partnerships with entities in the Outside dimension; along with ensuring that multiple feedback loops are in place so the museum can monitor public engagement and impacts.

What is perhaps most novel about the Inside-Outside Model is that it acknowledges that public cultural organisations are most effective when they respond to the changing trends and needs of the culture, in ways that generate adaptive impacts on that culture. This takes nothing away from collections and discipline-based expertise, but it does focus on impacts beyond those involving individual visitors.

Since humans first walked on Earth, culture has always been in a state of change. Such changes can be either adaptive (moving towards stability and balance) or maladaptive (moving towards instability and imbalance), within their ever-evolving contexts. Change has also been a characteristic of museums. However, museum change may be less focused on how they relate to the living culture and more focused on changes related to academic disciplines, collectors, markets, donors and funders. The idea of museums as catalysts of adaptive cultural change is relatively rare in the museum world, with the exception of ecomuseology.

It is important to acknowledge that museums have largely been instruments of colonial thinking and acting. It is widely known that many museums acquired collections that were taken from marginalised and/or oppressed people. Museum stories and histories have mistreated non-dominant cultures by omitting perspectives, erasing histories and by using stereotypes to perpetuate public misunderstandings and lies. Accordingly, when museums decide to embrace new potential public functions, like becoming catalysts of cultural change, it requires concerted effort to acknowledge, own and then dismantle residual elements of its own cultural past. Currently, many museums around the world have embarked on processes of rectifying racist parts of their own past. This is vital work in the Inside Dimension – and is necessary for museums to generate credibility as convenors and facilitators of public engagement on cultural issues.

It is important to add that these issues of systemic inequity continue to be deeply problematic within the living culture. As sustainability-engaged museums expand their commitment to addressing environmental crises, it is vital that they also address the social injustices, especially related to systemic inequity. Accordingly, the next section will address different facets of the Outside dimension.

The natural environment contains everything required to support human life – and humanity relies on it for its very existence. For that reason, the health and well-being of the environment should be of paramount concern for humanity. When the relatively stable balance of Earth's climate over recent millennia was knocked off kilter by the onset of the Anthropocene, all life that depends on natural systems must either adapt or deal with the consequences. So, in Figure 4.4, a small sampling of the human-derived threats to nature's complex systems is identified.

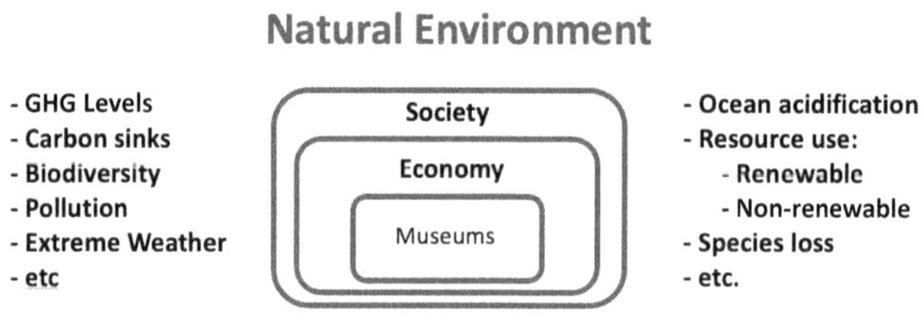

FIGURE 4.4 I-O Model – outside dimension

Pictured here, within the framework of Earth's natural environment, is human society, and, within that, the human-made economy. Museums are shown as being a subset contained within the economy and society, all of which relies on Nature. Since the deteriorating state of Nature's systems is being driven by humanity's outsized impacts, it is only changes to humanity's way of relating to Nature that can help reclaim some sense of relative balance. For addressing the cultural issues of our time, museums will need intelligence, creativity, compassion and leverage. Mobilising in this way will require courage.

However, it takes more than simply acknowledging this relationship to rescue it from the brink of planetary systems collapse (Diamond 2005). Humanity needs to grasp what scientists understand about rising greenhouse gas (GHG) emissions and the deadly levels that are approaching. For a museum to be engaged in reducing their own GHG emissions is not nearly enough. The promise of museums is not contained in a commitment to more efficient and less polluting versions of themselves. Rather the promise of museums is that they can become catalysts of cultural change across the entire living culture. And for that, a museum needs to know how it will monitor the essential feedback loops associated with the change it hopes to catalyse. At one level, this requires an understanding of the trends in global concentrations of greenhouse gas emissions. At another level, it means helping to ensure the public is aware, engaged and creatively active at dramatically reducing these emissions. Most of this will involve museums and the living culture, acquiring and privileging new skills, knowledge and behaviours.

Museum visitor studies have long experience of understanding motivations and demographics of audiences; however, they have struggled to understand the longer-term impact of museum engagement for individuals. New approaches are needed for museums to effectively support communities to address vital issues in ways that are relevant, build social cohesion and foster a shared vision of the future (Worts 2011).

It can be extremely helpful when museums understand how well their public engagement strategies actually have measurable impacts on individuals and conversely, how individual perspectives and experiences can have significant impacts on museums. Although there is a long list of possible impacts of museums on individuals, some of the core ones are given in Figure 4.5. Museums can aim to support individuals becoming inspired to understand the issues and forces that are shaping their culture, both intentionally and unintentionally. Such understanding can lead people to act in ways that foster well-being in themselves, their families, communities, cities, bioregions and social systems. These impacts can contribute to a healthy, engaged, democratic and sustainable culture.

When museums relate to people with respect, honesty, compassion and trust, members of the public can become more cohesive and motivated to engage in the living culture. It is not that museums should tell individuals what to think or do about the issues of the day, but rather their power is to invite the public into processes of reflection, discussion and action that are timely and relevant. This

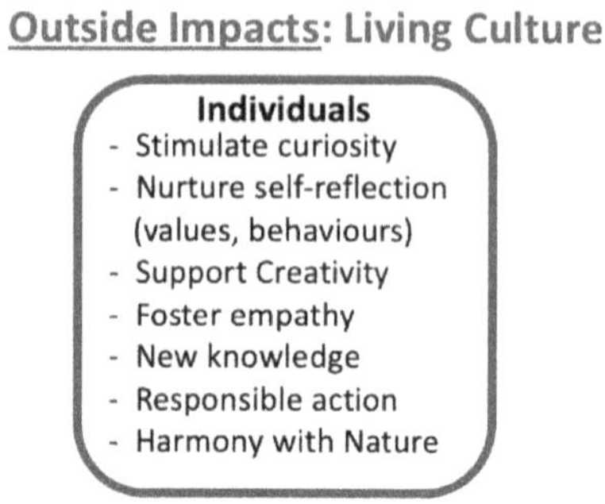

FIGURE 4.5 I-O Model – outside impacts – individuals

approach to museology was core to the early work of ecomuseums (de Varine 2017) and community museums (under the banner of the New Museology), but there are still few museums that are truly weaving together socio-economic and environmental trends and developing integrated approaches to imagining viable futures and building social cohesion, equity and catalysing co-creative action. In North America, museums are still solidly planted in the leisure-time economy, and the ideas of ecomuseology and the New Museology have done little to affect the traditional vision, structures and practices of museums.

Traditional museums are still often designed to welcome visitors who either live away, and happen to be visiting in the role of tourists, or appeal to local people who visit occasionally, often for a special exhibit or to entertain out-of-town visitors. In both cases, the opportunity to actually build ongoing relationships with these occasional visitors is extremely limited.

However, if museums could develop strategies that prioritise the building of relationships with local citizens, around contemporary issues, the potential for more cultural involvement and cohesion can be created. Museums could help support individuals, and groups for that matter, as they: engage with both historical and contemporary issues/materials; connect with wide-ranging visions of proposed futures (from the viable to the non-viable) and exchange perspectives with others about overlapping interests. Such activities can lead to new potential cultural impacts. But such an approach to facilitating new forms of cultural dynamics will require museums to experiment with new public involvement strategies – and assess how visitors actually engage (Worts 2016).

Beyond individuals, museums can connect with groups in meaningful ways. Individuals spend a lot of their lives in relationships with groups of one sort or another, including those related to a common heritage, special interests or a shared neighbourhood (see Figure 4.6). Perhaps the most common example of a group is the family. It is within families that many people learn the basics of how to interact with others, as they gain understanding of how to navigate the needs and opportunities presented by doing things with others. Most mainstream museums have

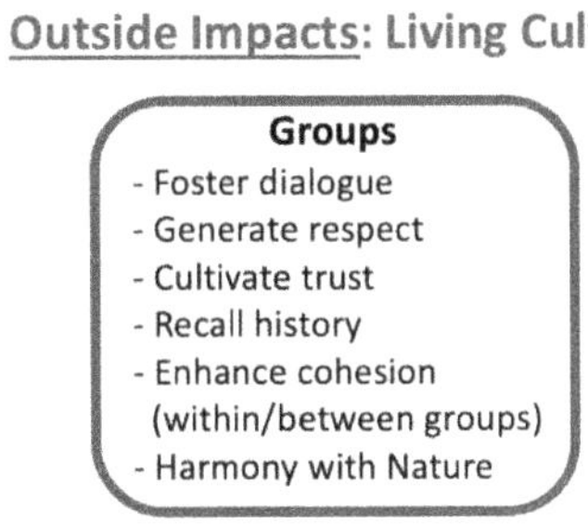

FIGURE 4.6 I-O Model – outside impacts – groups

already developed strategies to engage with families – for which there is an extensive museological literature.[7] A few family-oriented or other types of group-based museum programs have ventured into the sustainability realm, such as the Horniman, the Whitworth and Leeds Museums in the UK (see examples in this volume). However, it is a potentially fertile ground for opening up dialogues around issues of values-based decisions, the implications of scaling common practices, assumptions about the future we imagine we are headed towards and data on where current trends are actually taking us.

For groups to function well, there must be trust and respect and a sense of shared values. Interacting with historical topics and materials is a rich way for individuals to explore, understand and ultimately nurture shared visions of the future and an ethical way to live meaningful lives and more.

While groups are often brought together by meaningful common ground, communities and neighbourhoods are often characterised by some degree of common interest, but as often as not, considerable differences.

Most museums exist in local communities, but not all have strong relationships with them. And if museums always aim to define their relationship to a community within the context of people visiting the museum building, the relationship can be seen as lop-sided and more transactional in nature, rather than being relational. Since mainstream museums are built on the notion of audience transactions, as opposed to growing, evolving relationships, they often have difficulty expanding their reach beyond museum property. Figure 4.7 shows how museums can work to develop positive impacts on their communities. It is designed to signal that museums can (and perhaps even 'should') be expanding beyond individual and family focuses to develop novel impact and process goals related to cultural adaptation. While adaptive and non-adaptive processes have been part of human evolution always, as we pushed into the Anthropocene, we have unconsciously and foolishly ignored the signals telling humanity to change course. Museums have been almost entirely disengaged in such processes – because they have been operating in the culture without the right skills, visions and strategies for fostering adaptation. Museums therefore need to innovate to become influential in the dynamics of a

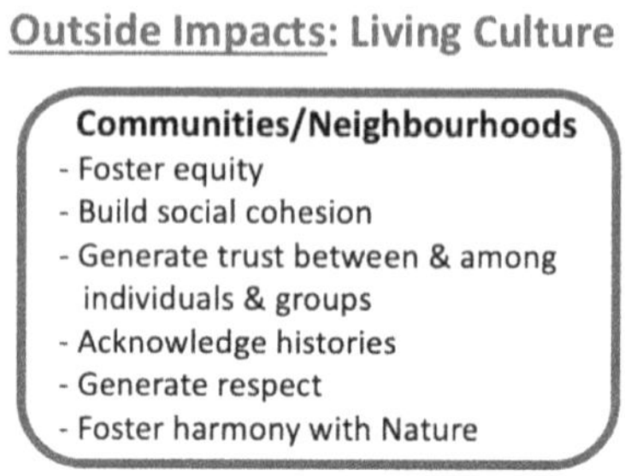

FIGURE 4.7 I-O Model – outside impacts – communities

culture that hopes to adapt and survive. Museums should be striving to create new kinds of impacts – that often will not be like 'museum impacts'.

One of the big questions that museums must grapple with is 'how can museums play the role of cultural catalyst, without being manipulative'? Another is 'how do museums support the creative interactions of elements making up communities, without making themselves an integral, ongoing part of the dynamic'? These are relatively new skills for museums, so much experimentation, assessment and adjustments will be necessary. Good examples have emerged particularly in Europe such as the Happy Museum Project[8] and the Museums Change Lives initiative of the UK's Museums Association (Museums Association 2013).

Organisations in general (including museums) are building blocks of societal systems (see Figure 4.8). Currently they play a wide range of roles, as: for-profits (free-market), non-profits, governments, educational systems and more. Organisations are designed to help achieve goals within cultures. For over a century, corporations have been given various sorts of powers through laws and conventions that are built on assumptions, principles and trust. For example for-profit corporations were historically designed to efficiently deliver a product or service. Examples include building and operating railways and generating energy, mining and manufacturing. Society envisioned a for-profit sector that effectively generated financial wealth as it delivered material goods and services and provided jobs. The non-profit sector was largely geared to helping society manage the unintended consequences produced by the for-profit sector (e.g. cleaning up negative environmental and social impacts that needed to be addressed) or addressing issues that the market would not support (education, humanitarian work, self-realisation). Charities offered a way to move money and services from those with money to those without. Government was designed to look after societal well-being (especially police, hospitals, schools) to ensure democratic governance, as well as address problems that were unanticipated (e.g. disasters).

But many organisations that work well at one point do not necessarily continue to do so over time, unless they adapt to changing circumstances. For-profit organizations that were designed to generate financial wealth, specifically through production

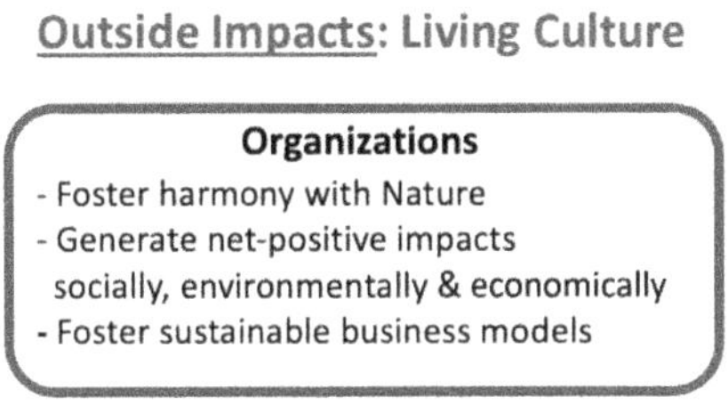

FIGURE 4.8 I-O Model – outside impacts – organisations

and consumption of goods and services, are now facing a rude awakening – the Anthropocene. So-called 'for-profit' organisations, for example, have long operated with a false sense that they pay the costs of doing business. However, historically, many costs have been externalised, like pollution and loss of biodiversity. Now, with the Planetary Boundaries exceeded, these organizations must be held to account. Also, systems of competition systematically produce inequality and need to be rethought (see e.g. Monbiot 2022; Raworth 2017; Thunberg 2022).

The point here is that status-quo organisations cannot be considered sacrosanct in a world that is fundamentally changing. If humanity's ultimate goal is to retain a healthy balance within planetary systems, over time, then the governing systems for organisations must always be part of the mix as adaptive change is being ever-cultivated. Building agreements on the overarching principles for the living culture, over time, is also part of the ongoing challenge. Museums can play important roles in such processes, because they can engage the public in thoughtful reflection, dialogue and co-creative action.

Cities and regions are made up of all the components discussed up to now, including the bioregions in which they are located. Cities/regions have a vital role to play in forging cultures that meet the needs of both present and futures (see Figure 4.9). In fact, they may become more vital than ever, because of the Anthropocene. Cities/regions are perhaps the largest or highest level of organisation that is capable of understanding, and relating to, all of the other levels – individuals, groups, organisations, communities, natural systems and more. All of the stakeholders that exist within a region need to be involved in the co-creative vision-building, strategising and actions that are designed to continuously build towards an evolving vision of the future, and museums can play a catalytic role in this.

As time goes on, there may be increasing pressure to organise human settlements around bioregions, because, in today's world, the vast majority of materials originate beyond the locality where they are consumed. Shipping goods and materials around the world is exceedingly problematic, not because of the monetary cost, but because of how our economic and business systems have 'externalized' so many real costs – leaving nobody accountable for the damage that is done. So, becoming food secure within bioregions makes a huge amount of sense. Governments that

Outside Impacts: Living Culture

Cities/Regions

- Food security	- Effective transportation
- Sustainable energy	- Live/work/play
- Equitable employment	- Participatory democracy
- Equitable education	- Fosters harmony with Natire

FIGURE 4.9 I-O Model – outside impacts – cities/regions

are organised to manage bioregions, not simply politically defined spaces we call cities/towns, may help to plan effectively for balanced approaches to environmentally/scientifically viable and ethically desirable human settlements.

It has been a common phenomenon for cities to experience exponential population growth, which necessitates the provision of ever-increasing housing, food and a host of services.[9] Much of the housing in our culture is considered a market commodity, and a vehicle for individuals and corporations to make huge amounts of money. Sadly, the pursuit of profit has devastated large amounts of prime agricultural land in the rush to build urban and suburban sprawl. With the loss of open land, both agricultural and 'wild', the human/Nature relationship is threatened. Local populations become more disconnected from a reliable source of food, since local farming is unable to produce sufficient food to meet local demand. The result of that is increased pressure on food production using high-intensity agricultural techniques that erode soil health and then shipping food around the world, with massive carbon footprints (Monbiot 2022; Rees and Wackernagel 1996).

Cultures that lose their ability to be adaptive in our fast-changing world risk having Nature rebalance its own systems, with no concern for the well-being of any particular species (i.e. humans). As a result, regional approaches to culture could help generate viable, shared visions of the future; monitor current trends and develop new strategies that ensure well-being for all stakeholders within a healthy, conscious and adaptive region.

Beyond the level of city/region, it is clear that national governments play an important role, especially if humanity is to be able to 'think globally and act locally'. It is important to remember, however, that governments that are most distant from their constituents are those at the national level. It makes a lot of sense to enable lower levels of government to address needs and opportunities within a region. National governments, at least in theory, exist to ensure that equity and well-being are foundational parts of a population that stretches over multiple regions. They also connect with and help to harmonise realities in other nations and parts of the world.

To reach one step higher and to imagine how global governance might better operate, it is worth looking at existing models. The United Nations is an example of how challenging it is to bring the world's countries together in an effort to agree on a common future. Through the Sustainable Development Goals (SDGs), a consensus plan was developed and agreed to in 2015.[10] Using an understanding that all of the world's systems are interdependent, the SDGs tease out 17 goals, which are both unique and entirely interdependent. Each country has agreed to (but is not legally bound to) address the SDGs in ways that are appropriate for that country. These goals are not a perfect solution, but rather a framework for each country to clarify the nature of the challenge/opportunity in that jurisdiction and develop its own approach to a sustainability vision and plan. Each country feels a sense of ownership for its challenges and solutions. The SDGs provide a useful tool to help guide cultural organisations towards meaningful cultural impacts (McGhie 2019).

The high level of systemic inequality (both social and economic) that exists around our globalised and interdependent world makes it difficult to design a future that treats everyone fairly and sustainably. Great economic wealth and power exist in some places and not others, all based on values, systems and behaviours that are not possible to sustain. As a result, it is vital that the foundation of an emerging, globalised future acknowledge and honour its multifaceted past. Equally important is that the future is based on a truly level-playing field based on equity, justice and living within the Planetary Boundaries. This brings us to perhaps the most challenging part of humanity's future – to transform systems that have evolved over millennia (see Figure 4.10).

Unless humanity can alter many of the systems associated with the accumulation of power and wealth, it is hard to imagine how there is a future for humans on Earth – certainly not a future of well-being. Essential for well-being is that we live within the biophysical limits of the planet. And arguably, we cannot continue without systems that ensure equity for all. This means that we need economic and governance systems that are designed to produce these results. There are no quick fixes for these things, but they stand as the backdrop for all that museums might do in their efforts to foster a global culture of sustainability.

In order to bring us back down from the stratosphere of puzzling over how best to design sustainable global systems, it is worth returning to the inside dimension of the Inside-Outside Model (see Figure 4.11). It is here that we must remember that if museums are to become catalysts of cultural adaptation and transformation,

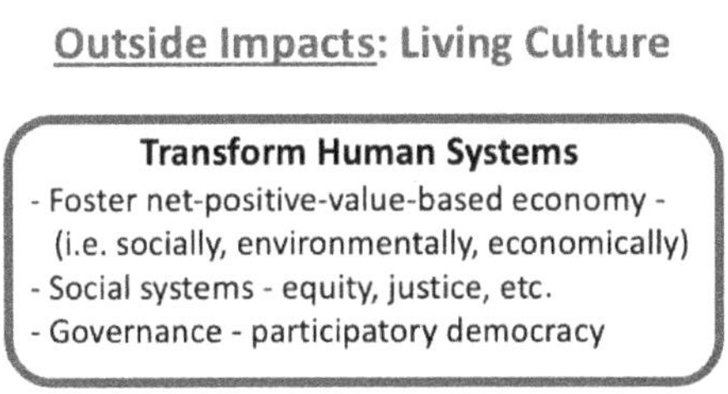

FIGURE 4.10 I-O Model – outside impacts – human systems

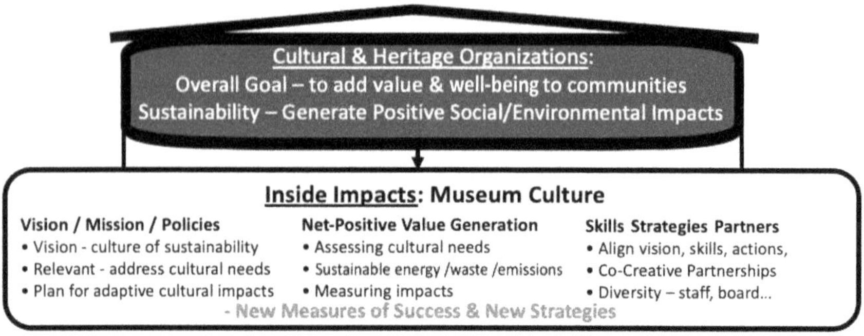

FIGURE 4.11 I-O Model – inside dimension *revisited*

they will need to create 'New Public Engagement Strategies' and 'New Measures of Success'. These are natural by-products of thinking more holistically. It is the only way we can break out of the cycle of doing what we've always done.

The Inside-Outside Model is a relatively simple tool that was designed to help map a very complex set of dynamics related to the living culture, sustainability and museums/ecomuseums. It does not contain answers to the question of 'what should museums do to have meaningful cultural impacts?'; however, it does offer a framework for designing public engagement strategies that have the ability to catalyse such impacts.

Two notions have emerged through the development and working of this model over the past few years. The first is how valuable it is if museums create a 'theory of change' that would help guide each of its public dimension activities. By considering and then articulating what museums hope is different as a result of their public engagement, they align the desired impacts and the strategies they plan to use to foster them.[11] Since sustainability and unsustainability are understood as characteristics of large, complex, interdependent systems, and not simply local strategies of operational efficiency, museums could plan for co-creative work with many societal stakeholders in processes of cultural transformation. This approach will required additional skills to those used to deliver cultural experiences, usually within the leisure time economy, but such skills are those needed for cultural adaptation.

The second notion that emerged from this work on the I-O Model are the potential benefits of approaching cultural planning as a process of value-generation. This simply means becoming clear about how both positive and negative values are generated. The positive side is likely to involve intentionally facilitating: (1) deep reflection; (2) dialogue and cohesion-building; as well as, (3) co-creative action – which can apply to many different stakeholder categories (including individuals, groups, organizations, etc.). The negative value that is generated is usually unintentional side effects of 'business as usual'. Both need to be woven together in planning processes that bring conscious awareness and responsible action to a central role in museum planning and cultural life.

Cultural adaptation is what humanity needs to ensure a future for humanity. It will require engaging the potential vision and creativity and that lies at the heart of our species. Museums, and their insights into the creative muses, have great untapped potential. They need to decide if they have the courage to step into the role of cultural catalyst.

Ultimately, Individual museums cannot single-handedly take on the culture of unsustainability that humanity has created. It requires that the field has some shared sense of purpose regarding catalysing cultural adaptation. It also requires that our museology evolves beyond the traditional pillars of collections, exhibits and on-site programs. And it requires new models of enabling this work that is not reliant on grants from governments and businesses – which too often come with the imperative that uncomfortable topics be checked at the door.

If sustainability is in humanity's future, all other sectors need to change as well – business, governance, education and more. So, figuring out how to transform the economy into a force that does not 'dig the hole deeper' is not for museums to figure out – but they can help convene and facilitate the processes in ways that are not part of leisure-time entertainment but rather tending to the cultural well-being of people. As Daniel Christian Wahl has long called for, regenerative well-being of both planet and people must form the foundation of a future vision of our culture (Wahl 2016). Museums can become catalysts of forging the vision – and supporting its realization.

Notes

1 This chapter has evolved from an online presentation, delivered by Douglas Worts, at the 'Ecomuseums and Climate Action Conference' held on 30 September 2021, which was part of the Pre-COP26 Meeting in Milan, and from the two subsequent articles by Douglas Worts and Raul Dal Santo, for the conference book *Ecomuseums and Climate Change* (2023).
2 See https://en.wikipedia.org/wiki/Holocene
3 Planetary Boundaries refer to a set of Earth's natural systems that are vital for the stability of our planet – developed at the Stockholm Resilience Institute, in 2009. https://en.wikipedia.org/wiki/Planetary_boundaries
4 'Doughnut Economics' is a macroeconomic framework, developed by Kate Raworth, a UK economist, who published a book with the same name (2017).
5 Doughnut Economics Action Lab (DEAL)– see https://doughnuteconomics.org/
6 For example the B Corp Movement – https://bcorporation.eu/country_partner/italy/
7 See, for example, the Association of Children's Museums Research Network –https://childrensmuseums.org/initiatives/childrens-museums-research-network/
8 https://happymuseumproject.org/
9 https://www.worldbank.org/en/topic/urbandevelopment/overview
10 https://sdgs.un.org/goals
11 See https://www.museumsassociation.org/campaigns/museums-change-lives/measuring-socially-engaged-practice/getting-started/ or https://www.artsculturefinance.org/wp-content/uploads/2017/09/Arts-Impact-Fund-Guide-Template-for-a-Theory-of-Change.pdf

References

AtKisson, A. 2010. *Believing Cassandra: How to Be an Optimist in a Pessimist's World.* London: Routledge.

Brundtland, G.H. 1987. *Our Common Future: Report of the World Commission on Environment and Development.* Geneva: United Nations.

Dal Santo, R. and Worts, D. 2023. *Museums Planning for Cultural Impacts: The Case Study of Parabiago Ecomuseum (Italy), in Ecomuseums and Climate Change.* Italy: Ledi Publishing. https://www.ledipublishing.com/book/9788855268387/ecomuseums-and-climate-change

Davis, P. 2011. *Ecomuseums: A Sense of Place.* London: Continuum.

De Varine, H. 2017. *L'écomusée singulier et pluriel: Un Témoignage Sur Cinquante Ans De Muséologie Communautaire Dans Le Monde.* Paris: L'Harmattan.

Diamond, J. 2005. *Collapse: How Societies Choose to Fail or Succeed.* London: Viking.

DROPS Website, the World Platform for Ecomuseums and Community Museums. https://sites.google.com/view/drops-platform/home

Hirzy, E. (ed). 1992. *Excellence and Equity: Education and the Public Dimension of Museums.* Washington: American Alliance of Museums.

Homer-Dixon, T. 2006. *The Up-Side of Down: Catastrophe, Creativity and the Renewal of Civilization.* Toronto: Vintage Canada.

Janes, R. 2009. *Museums in a Troubled World.* London: Routledge.

Klomp, K. and Oosterwaal, S. 2021. *Thrive: Fundamentals of a New Economy.* Uitgeverij: Atlas Contact.

Koster, E. 2020. Relevance of Museums to the Anthropocene. *Informal Learning Review,* 161: 18–34.

McGhie, H. 2019. *Museums and the Sustainable Development Goals.* Curating Tomorrow. http://www.curatingtomorrow.co.uk/wp-content/uploads/2020/01/museums-and-the-sustainable-development-goals-2019.pdf (Accessed 8 October 2022).

Meadows, D., Meadows, D. and Randers, J. 2004. *Limits to Growth – The Thirty Year Update.* Vermont: Chelsea Green Publishing.

Meadows, D., Meadows, D., Randers, J. and. Behrens III, W. 1972. *Limits to Growth.* New York: Universe Books.

Monbiot, G. 2022. *Regenesis: Feeding the World Without Devouring the Planet.* London: Allen Lane.

Museums Association. 2013. *Museums Change Lives: The MA's Vision for The Impact of Museums.* London: Museums Association.

Raworth, K. 2017. *Doughnut Economics: Seven Ways to Think Like a 21st-Century Economist.* London: Random House.

Rees, W. and Wackernagel, M. 1996. *Our Ecological Footprint: Reducing Human Impact on the Earth.* Philadelphia: New Society Publishers.

Riva, R. (ed). 2017. *Ecomuseums and Cultural Landscapes: State of the Art and Future Prospects.* Milan: Maggioli Editore.

Riva, R., Aldovini, G. and Dal Santo, R. 2022. Managing the Eco-Social Transition: Communities in Action to Build Possible Futures. *TECHNE: Journal of Technology for Architecture and Environment,* 23: 62–68.

Sutter, G. and Worts, D. 2005. Negotiating a Sustainable Path: Museums & Societal Therapy. In Janes, R. and Contay, G.C. (eds) *Looking Reality in the Eye: Museums & Social Responsibility.* Calgary: University of Calgary Press, 129–152.

Thunberg, G. 2022. *The Climate Book.* London: Allen Lane.

Wahl, D.C. 2016. *Designing Regenerative Cultures*. Axminster: Triarchy Press.

Worts, D. 2003. On the Brink of Irrelevance? Art Museums in Contemporary Society. In Tickle, L. and Xanthoudaki, M. (eds) *Researching Visual Arts Education in Museums and Galleries: An International Reader*. Dordrecht: Kluwer Publishers, 215–231.

Worts, D. 2011. Culture and Museums in the Winds of Change: The Need for Cultural Indicators. *Journal of Culture and Local Governance*, 3(1–2). https://uottawa.scholarsportal.info/ojs/index.php/clg-cgl/article/view/190/173

Worts, D. 2016. Museums: Fostering a Culture of 'Flourishing'. *Curator*, 59(3): 209–218.

Worts, D. 2017. Planning for Cultural Relevance: A Systems Workshop at the Georgia O'Keeffe Museum. In Yung, J. and Love, A.R. (eds) *Systems Thinking in Museums: Theory and Practice*. Maryland: Roman & Littlefield, 81–93.

Worts, D. 2019. Heritage Planning for Sustainable Cultural Impacts. *AASLH Blog Post*. https://aaslh.org/sustainable-cultural-impacts/ (Accessed 12 September 2022).

Worts, D. and Dal Santo, R. 2023. The Inside-Outside Model – Animating the Muses for Cultural Transformation Amid the Climate Crisis. In Davis, P., Borelli, N. and Dal Santo, R. (eds) *Ecomuseums and Climate Change*. Milan: Ledi Publishing, 65–90.

5

TACKLING THE CLIMATE CRISIS

An overview of UK museums

Kathryn Simpson

Introduction

As part of a widespread movement that grew from grassroots, museums added their voices to global declarations of a climate emergency in 2019. While many joining the call didn't yet have detailed plans on how their organisations would tackle it, these commitments signalled the start of a more concerted effort to drive change.

The imperative to reach net zero has never been clearer. As of 2019, some museums were already making good progress to decarbonise and had well-established sustainability plans. In fact, many of the museums that were further ahead have also since committed to science-based targets to reach net zero much sooner than Government targets. For other museums, this moment signalled the beginning of a green transition, and they are now in the process of establishing how to reduce emissions on the necessary scale. Despite some making great strides and a widespread sense that things are now beginning to change, the sector is still relatively immature on its journey to reach net zero or other environmental goals, but that's not for a lack of ambition or determination.

The obstacles that face the sector have made change difficult, but not impossible. As outlined in this chapter, the efforts of museums and the dedicated people who work in them are starting to create momentum; the state-reliant and often precarious funding models, long-standing issues with infrastructure and estates and the barriers created by the types of buildings and assets museums occupy and care for are beginning to be addressed.

The pace of change must accelerate, and there is broad agreement that to do so the whole sector needs to work together to address the unique complications that museums face.

DOI: 10.4324/9781003347606-6

So, what does it mean in practice for museums to commit to tackling the climate emergency? Where can they make an impact and how far have they come? This chapter will highlight the actions museums are taking to reduce energy consumption and how they're using collections to further development of essential scientific research and supporting the public to learn about and interrogate the impact of their activities and those of others.

Developing a plan

Though making public statements is a positive move, it doesn't necessarily deliver action. However, in many cases communicating the will to take action publicly was a pivotal first step for making organisation-wide commitments to action and identifying the practical steps required to deliver. To do this work, museums needed to engage their teams, public and funders with their aims to make the green transition a reality. For many, this was a major part of achieving a shifting of position, and so this strategic reprioritising was helped along by a public commitment.

Although climate adaptation is something many sectors have been developing for years, overall, museums have been comparatively slow on the uptake. The reasons for this are myriad and include but are not limited to: reliance on public, grant and philanthropic funding; long-standing issues with core funding; unaddressed issues with ageing and failing infrastructure and estates and museum collections' specific challenges which require a duty of care to objects and conservation. The practical and physical barriers have also been coupled with philosophical and political barriers, from risk management and aversion, fear of lacking expertise in some key areas and a lack of funding and public or political backing.

As most museums are in receipt of public funding, some of it directly or at arms-length from government, the government's commitment in 2019 to reach net zero by 2050 heralded something of a turning point for museums feeling they could more proactively respond to the crisis. They now have the impetus and crucially the public backing (Ipsos MORI Political Monitor 2022) to put their long-held plans and ambitions into action.

Getting on the road to environmental sustainability does come with challenges, and where expertise is missing or institutional knowledge nascent in the sector, museums have needed to look to others to begin. Some trailblazers have set the pace whereas others have struggled to make inroads, often reporting feeling daunted about a lack of capacity and expertise. But this problem of where to begin is at least shared, and so a primary focus of the collaborative work in the sector so far has been on sharing information and best practice tips and tools to enable learning from one another.

The difficulties in moving from announcement to action are recognised to the extent that local government commissioned a guide aptly titled *You've declared a Climate Emergency, what now?* (Arup 2019). Networks, conferences and workshops have often focused on initiatives outside of the UK where more progress has been made in certain areas. In Australia, for example many communities have more readily felt the impacts of climate change, and so museums such as the Australian Museum have responded to these challenges and shared details of their journey and helpful insight with the wider international museum community (see Newell and Ahmed, this volume; Newell et al. 2017).

Museums have moved from recognising the need for action, and asserting that need publicly, to identifying how to deliver across their organisations, while taking audiences and funders with them. They are taking a wide range of approaches, with some more developed than others. Many are creating sustainability plans to complement existing strategies, while others are incorporating environmental goals within wider organisational plans. Bespoke sustainability plans can help organisations to outline the exact steps needed to create and deliver change, and they can also be costed where museums have capacity to do this, while embedding sustainability throughout organisational strategies makes the need for change an integral part of the museum's delivery. Some strategies start small from the needs of the organisation and immediate audience, and others look beyond to a wider role in the community or as global players in reaching sustainability aims such as Royal Botanical Gardens Kew and the Natural History Museum.[1]

The V&A's sustainability plan (V&A 2021) targets net zero emissions by 2035 and strategically links to the museum's overall mission with an ambition to 'enable people, through art, design and performance, to contribute to a thriving planet and shape sustainable futures'. A key element of the plan is to harness the 'enthusiasm, creativity and expertise of . . . employees and volunteers to embed sustainable practice across the organisation'. One of the ways in which it does this is through a sustainability-focused learning programme, throughout which employees and volunteers are offered the opportunity to engage in a range of topics, starting with carbon literacy and moving through finance and inequality, sustainable materials and intersectional sustainability. The museum used a process of outcome-based learning, seeking continual feedback on the future shape of the programme: by equipping and empowering employees in this way, the tools and knowledge to reach the organisation's sustainability goals are shared and strengthened.

The V&A also has an 'Anthropocene Reading Group' which is a cross-department discussion forum dedicated to critical thinking that brings together museum staff with external organisations from multiple disciplines to interrogate concepts and theoretical frameworks and identify opportunities for research, programming and digital initiatives to progress sustainability ideas.

Royal Botanic Gardens Kew (RBGK) holds the world's largest and most diverse botanical and mycological collections with over 8.5 million items and has committed to an ambitious programme to refocus all of its work on ending biodiversity loss and mitigating the impacts of climate change. The organisation is working towards the goals of three plans that will operate in tandem: a corporate strategy 'Our Manifesto for Change 2021–2030' (RBGK 2021a) to end the extinction crisis and protect nature; scientific priorities in the 'Kew Science Strategy 2021–25' (RBGK 2021b) which seek transformative change and maximum impact and the 'Sustainability Strategy: urgently tackling the climate and biodiversity crisis' (RBGK 2021c).

Planet PK is a programme developed by PK Porthcurno – the Museum of Global Communications which embeds green values in every decision it makes from business planning to programming and marketing. Planet PK has not only won awards for its own on-site activities and approach to sustainability but also used its position to partner with others to make change (PK Porthcurno 2022). Planet PK is a 'Plastic Free Business Champion' with Surfers Against Sewage (SAS), a partnership which has led to the development of a Plastic Free Steering Group for the whole of Porthcurno. The museum also played a part in successfully securing West Penwith's status as an International Dark Sky Park as a member of the project steering group and hosting a thematic exhibition to raise public awareness of light pollution.

The Natural History Museum (NHM) also takes an organisation-wide approach to the planetary emergency with its central strategy to 2031 (NHM 2020) setting out the role the museum will play as a 'global, scientific and cultural leader' and with a stated mission to 'create climate advocates for the planet' (see Matterson, this volume).

This approach of using museum collections to educate and equip the public with the knowledge and inspiration to make changes is a major theme of how museums engage the public which is explored in the next section.

Engaging the public

Museums and their collections are an obvious place for discussions about climate and biodiversity to take place. They can also link up the interconnected conversation around social and climate justice as many museums' histories and roots can tell stories about industrialisation and colonialism.

A broad shift has taken place inside and outside the museum sector regarding what they do and what they're for. Debates have moved from a view that museums are purely object-oriented 'neutral spaces' to more issue-focused organisations. Museums have realised that they have a voice through which they can engage the public and assert values for positive change, that museums are about people and the values they hold and that the 'museum voice' can be active, engaged and challenged in debates.

Engagement with the public on these issues can take many forms, from informing, activating, inviting in, challenging or with the institution itself being challenged. Museums have taken the spotlight on many issues concerning social justice and are increasingly sites of protest and debate. As cornerstones of civil society with high levels of public trust (Ipsos 2021; Collins 2021), museums must take their role seriously when it comes to approaching topics of this nature, especially when many have grown out of a legacy of colonialism or as founding monuments to industrialisation and thus contributed to the changing climate.

This section looks at the ways that UK museums are working with their communities to address climate and biodiversity issues. As places that seek to inspire and educate, museums can't only present the devastating impacts of climate change. Alongside exploring these very real topics, they must also seek to engage the public through action-oriented exhibitions to portray a more hopeful future.

Museums have access to large public networks and have developed deep links with their audiences through mass community participation projects such as the Royal Albert Memorial Museum and Art Gallery's (RAMM's) *Growing Together* which used seed planting and growing to improve community well-being in Exeter (NMDC 2021). Following a long winter lockdown, the museum sent out seed packs to 50 community organisations in the lead up to its summer exhibition *Seedscapes: Future-Proofing Nature* and *Léonie Hampton: A Language of Seeds*. The packs encouraged participants to plant one species a week and learn facts along the way (RAMM 2022).

Museums Partnership Reading (Reading Museum and the Museum of English Rural Life) has a burgeoning sustainability initiative called *Our Green Stories; sustainability in our communities*. Using the UN's Sustainable Development Goals and guidance from Arts Council England (ACE) as inspiration, it uses public engagement and research to tell green stories so the public can read and engage. It combines history with wildlife and information from the collections and offers family events. The partnership defines sustainability locally and internationally and asks the public how much they know about green issues and the environment and where they feel they could make a difference.

Museums can nurture and inspire their audiences with projects about the environment, but some have also used shock tactics to communicate a message. The team at the Horniman made a surprising intervention with *Beat Plastic Pollution* which presented visitors with the threat facing the oceans. As Destre and Merriman describe in this volume, the visceral pop-up display inserted more than 150 items of single-use and waste plastic for several months into the permanent aquarium exhibits, prompting emotional responses in visitors.

For Oxford University Museum of Natural History's exhibition *Meat the Future* which explored the environmental impact of meat consumption, the team utilised the museum's role as a 'trusted voice' (British Council 2019) to present information about food production and environmental impact in an inspiring and surprising way to encourage visitors to make their own informed decisions about diet. Evaluation of the exhibition showed that 59% of visitors not only wanted to change the balance of their diet as a result but also commented that they appreciated the way information was presented as it didn't dictate particular actions (Museum of Natural History Oxford 2021). The museum's café *Eat the Future* created a pioneering partnership whereby visitors were invited to select their lunch based on informed choices via ecolabels.

Wakefield Museum's *A World of Good* immersive exhibition (which was shortlisted for the Museum and Heritage awards 2022 Sustainability Project of the Year and 2021 Best Project award) is a good example of activating the public to engage with climate issues, as the multi-sensory immersive exhibition asked visitors to pledge to change just one thing in their lives. An interactive website allows visitors to sign up to a range of actions to take in the home or garden and at work or lifestyle and invites people to make a pledge and spread the word. There are also a range of resources for teachers and schools including how to conduct species surveys and use a carbon calculator.

In *Inheriting the Anthropocene,* a project with 11–14-year-olds, Manchester Metropolitan University's Biosocial Research Lab and Manchester Museum showed the value of opening up the museum in a different way: by 'reimagining it and challenging some of the practices and ideas that often go unquestioned, they were able to rework the relations between people and things'. Gathering and incorporating the perspectives of young people into museum work can expose startling and motivating ideas, such as one young person who when asked what it felt like to be 'inheriting the Anthropocene' responded 'it's like being given the Mona Lisa and then realising someone's drawn a moustache on it' (NMDC 2021).

In Scarborough, the free exhibition *Dear plants . . . (Crescent Arts 2022)* was created in conjunction with young people, which opened with

> *[D]ear plants . . . is not a statement, but a sentence, one that needs adding to, but not ending, it is a call to look, listen and learn in order to create a more positive climate future and relationship with ourselves, plant and animal life.*

The project and exhibition were a collaboration with Crescent Arts Blueberry Academy, which supports young people and adults with learning differences to promote employability and independence, and artist gardener Meg Woodward-Hay at Woodend Galleries and Studios. It created an opportunity not only to share with the public what has been learned but also to collect plant knowledge from visitors.

With *Story:web*, Tyne and Wear Archives and Museums (TWAM) also chose to open the museum doors to outside perspectives by giving objects and their stories out to the world in the form of big data and offering the opportunity for anyone to imagine and interpret their stories. It asked 'what if we could create and share our own climate change stories?' and 'what if we blew the doors off museums releasing their objects and other stories into the world as big data for everyone to own?' because 'there is no single climate change story, every story has the power to inspire climate action' (Story:web 2022; NMDC 2021).

Beyond their immediate audiences, museums' public activities can also extend to more globally based advocacy and engagement. COP26 and COP27 saw many museums and cultural organisations making displays, exhibitions and events alongside the meeting of global leaders. *Story:web* was one of eight finalists to be selected to exhibit this idea of a 'museum of the future' at Glasgow Science Centre ahead of COP26.

The Natural History Museum has a regular presence at these global meetings, where it presents cutting edge science and uses expert educators to talk to those visiting about climate impacts and supports the voices of young climate activists and artists with their platforms. For the United Nations Biodiversity Conference COP15 in 2022, the museum supported ecological artist Thijs Biersteker to create a sculpture for the main exhibition hall in Montreal. *Econario* (2022) is a sculpture that grows, morphs or shrinks based on real-time policy decisions and the calculated impact on biodiversity, which are informed by the museum's PREDICTS database which includes 4.9 m data points.

Away from the COP events, artists are regularly supported by museums to tell the story of climate change and highlight the events taking place, such as Olafur Eliasson's *In Real Life* exhibition at Tate Modern, which simultaneously created a sense of wonder with a serious message about sustainability and biodiversity, aiming to stimulate debate. Eliasson's exhibition followed two previous iconic interventions which caught the public imagination, the *Weather Project* in 2003 and *Ice Watch* in 2018, which literally presented Greenland ice boulders shrinking in the sun outside the museum.

Manchester Museum's *Thinking Through Extinction* project bridges public engagement and research. The collaborative project invited artist Laurence Payot to explore the emotional impact of extinction and ask what it actually means for people from different backgrounds and experiences when faced with the reality in communications and public spaces. The programme was originally intended to take place in the museum, but due to COVID-19 was delivered through a series of online workshops and used creative writing, choreography, animation and music to engage over 150 people with the topic of extinction (NMDC 2021).

What is apparent is that the role of museums is shifting, and the climate and biodiversity challenge is something they can all support and further through large networks, deep engagement, knowledge and expertise in science. The great treasures of museum collections can help reveal and even combat climate and biodiversity threats; this is what we look at in the next section.

Research

Research is a fundamental activity of museums. It is the duty of curators to learn as much as possible about the objects in the collection and share this knowledge in a dialogue with audiences. The knowledge of the museum is often added to or challenged by audiences and communities. Wider research in science, technology and innovation has long been a mainstay of the sector, and there is huge potential in the UK's collections to learn and experiment with materials, to learn from nature and historic design techniques and to better understand the climate and environment through specimens and texts.

In recent years, UK Research and Innovation (UKRI), the UK Government funder of innovation, has come to recognise the potential for galleries, libraries, archival and museum collections as 'research infrastructure' and is working with these sectors on a range of initiatives to nurture and build research potential, including the development of infrastructure for conservation and heritage science to become linked up, shared and made more accessible. Many museums are involved in scientific and wider research partnerships, and a huge array of projects are focused on climate and biodiversity.

Much of this work takes place as part of cross-sector partnerships bringing together curators, scientists and academics from museums, universities and research organisations across the globe. These projects focus on a range of issues facing the planet, from protecting ocean ecosystems to understanding biodiversity and preventing further extinctions and ecological disaster.

Museums are increasingly using contemporary research and studies to engage the public with real-time science. The University Museum of Zoology in Cambridge is one organisation using specimen data to guide restoration work on nature reserves and engaging the public in the process to highlight why conservation efforts are needed. Museum butterfly specimen data are helping to reveal a clearer understanding of the wildlife that historically populated the Great Fen, a wetland where 99% of habitat has been lost since the 17th century. The museum holds a collection of diaries of 19th-century naturalist Leonard Jenyns who made notes on local butterfly species between 1820 and 1849. The project looks at the species that used to be present in the fenland and how they altered due to the changing environment of the area. It uses wildlife of the past to guide future conservation efforts in the area and is working with wildlife trusts to create a baseline from which to measure conservation initiatives (University of Cambridge 2022).

Similarly, Manchester Museum is the engagement partner in a three-year NERC (Natural Environment Research Council)-funded project to study the mass extinction caused by a meteor impact in the Gulf of Mexico 66 million years ago. Scientists at the universities of Manchester and Plymouth are studying the composition of fossil molecules from bacteria that lived in peat accumulating at the time of the meteor and looking at whether the temperature change in that period can be reconstructed. They are then comparing temperature rises currently being experienced to better predict Earth's future. The research was central to the new dinosaurs' display when the Museum reopened in 2023 (Manchester Museum n.d.; NMDC 2021).

National Museums Scotland's National Museums Collections Centre (NMCC) in North Edinburgh houses ten million natural science specimens and welcomes visitors from around the world to use the collections to better understand all aspects of the changing natural environment. As well as engaging local residents in making a positive impact on biodiversity, the Centre is conducting a base-level survey of biodiversity on its site and demonstrating the role of modest interventions. This long-term scheme will continue as the wider Granton Waterfront is developed (NMDC 2021).

Plants Under Pressure measures how many plant species are threatened with extinction, where these plants grow and why they are threatened, assessing the status and traits of thousands of species. Together with Royal Botanical Gardens Kew, NHM has produced the first IUCN (International Union for the Conservation of Nature) Sampled Red List Index (SRLI) for Plants for the Convention on Biological Diversity, and this work has fed into the recent report by the Inter-governmental Science Policy Platform on Biodiversity and Ecosystem Services (IPBES) to estimate that one million plant and animal species worldwide are threatened with extinction due to human activity (NMDC 2021).

NHM's *Reef Refugia* also funded by NERC is a joint investigation between the museum, UK university research partners from Cardiff and Exeter and the University of Malaysia Sabah. The findings of the project have contributed to policy change in Malaysian Borneo to develop a marine-protected area. The research looks at the conditions in which some coral reefs thrive against the odds and draws on studies of NHM's extensive coral collection including those collected by Charles Darwin on the voyage of the Beagle in the 1830s and ancient fossil corals from the West Midlands dating back 440 million years (NMDC 2021; UKRI n.d.). The research is continuing after receiving Horizon 2020 Excellent Science funding, a source of funding which is highly sought but difficult for most museums to apply for.

NHM employs some 350 scientists and is a behemoth when it comes to research activity. Its status as an Independent Research Organisation (IRO), which demonstrates its capacity to conduct in-house research and the ability to undertake and lead research programmes held by some (mostly national) museums, makes it eligible for funding via the UK's research councils and therefore increases the institution's ability to take part in and lead collaborative research all the more possible.

The NHM and other museums are also playing a part in global research efforts by sending their collections to teams in other parts of the world. In 2019/20, national museums sponsored by DCMS lent just under 500,000 specimens and samples to institutions across the globe, and 99% of these were from NHM, so its contribution to global research across a range of disciplines in arts and sciences is impressive.[2] For example, a set of 9,435 specimens of Afrotropical beetle were 'loaned to experts for identification, supporting their research, and improving knowledge of African fauna, including describing new species and providing baseline data to assess environmental and global change through time' (DCMS 2021).

It's unfortunate that IRO status is only held by a few museums, making this type of research funding largely out of reach to much of the museum sector, including counterintuitively many university museums (which under current criteria cannot be involved in bids by their own university) where there is enormous capacity to make a big impact through extensive networks and research capabilities. There is great untapped potential in the museum sector to further research and innovation in the environmental field, and in the coming years the hope is that this funding will become more widely available.

Smaller museums do however contribute to globally significant research, the Horniman, for example, became the first institution in the world to breed coral in captivity in 2013 (see Destre and Merriman, this volume). The museum's purpose-built aquarium laboratories use microprocessor technologies to analyse a range of cycles, so working with researchers from the University of Derby, they were able to replicate the environmental conditions for reef spawning. The team is now able to synchronise spawning with the parent colony, and the Florida Aquarium is using the same techniques to reproduce coral on a much larger scale. Project Coral is a prime example of a small institution with big aims, an approach that was recognised by the Art Fund in awarding it the title Museum of the Year in 2022 for its broader work on climate and engagement.

Estates and operations

Motivating change won't just come through engaging the public with research, exhibitions and programmes; museums must also outline and meet their own sustainability goals to demonstrate commitments to reducing impact. Changes to the way museums operate and manage estates demonstrate openness and transparency about this work and can inspire visitors and partners to assess the environmental impact of their own lives.

The buildings that they own or manage are a unique privilege for museums, but they also come with inherent problems. Challenges specific to the sector arise from the types of buildings they occupy and the collections they care for. Their locally, nationally and internationally significant collections make museums unique among other cultural organisations, but collections' care, display, movement and touring require certain conditions that also make them incredibly energy intensive. Museums aren't just spaces to hold collections, they are also places of work and active centres for visitors, so collections care has to also be balanced with these competing needs.

Arts Council England's annual report on the environmental performance of the organisations it funds shows that museums use more gas and electricity than all other art forms (Julie's Bicycle 2022). While many in the sector are working to change the accepted parameters for environmental conditions to reduce energy usage and environmental impact, progress until now has been slow, despite various efforts over the years to ignite change. The public ownership of many collections of national significance means that the appetite for risk in managing them has been low; however, attitudes are now changing, and much work is in progress to move to new approaches. These very specific collections-based challenges and the activity underway to address them are also explored in other chapters (e.g. Balshaw, this volume, Southwick, this volume).

From listed buildings and sprawling municipal centres to castles, ships, preserved mining villages, historic houses and aircraft hangars, museum buildings come in widely different shapes and sizes. While there are some new and relatively modern buildings, many museums are housed in either heritage buildings (often listed) or older municipal buildings in desperate need of maintenance. Planning rules about what changes can be made to heritage buildings can be prohibitive and make alterations difficult and costly while limiting the extent to which buildings can decarbonise. Funding for maintenance and repair has long been a challenge, and the sector has suffered from underinvestment for many years which makes refurbishments and retrofit even more complex or financially out of reach.

However, there are many organisations that are overcoming these challenges to repair, improve and adapt their buildings and operations to reduce energy use.

The Science Museum Group's (SMG's) ambitious Net Zero target commits its family of five museums to achieve science-based targets by 2033 (SMG 2021a). This includes the transformation of the estate at the Science & Industry Museum in Manchester with a £4.3 million government grant, where £2.6 million will transform the iconic Power Hall which showcases Europe's largest collection of working steam engines by installing roof insulation, glazing, a new electric boiler and a ground source heat pump (SMG 2021b). The project will reduce CO_2 emissions by 60% (515 tonnes) and improve energy efficiency. The museum group hopes the gallery will be a landmark of future green energy.

The Science Museum Group has also planted 44,000 native locally sourced trees at the National Collection Centre in Wiltshire where the majority of its collections are stored. The planting scheme is accompanied by 88 acres of photovoltaic arrays that generate about 50 GWh, enough energy to power over 15,400 homes annually.

The National Archives has 16 on-site repositories storing 168 km of physical documents, and effectively managing the repository environments is its greatest source of carbon emissions and energy consumption. Over 15 years, the Estates and Collection Care Departments have transformed the efficiency of the environmental management systems and driven new standards for the storage of library and archive collections nationally and internationally: PAS 198 (BSI 2012a), PD 5454 (BSI 2012b) and CEN EN 16893 (European Standards 2018). Through monitoring and pioneering research, the 24/7 air conditioning operation has been replaced by a demand-led environmental management system. The new operation involved powering down overnight and on non-working days. These new standards have encouraged energy-saving collections management across the sector and worldwide (NMDC 2021).

The Museum of Making at Derby Silk Mill has seen a decade in development and has been an innovative project since its inception, from its procurement model to the participatory process through which it was built with and for the community. A key objective of the development was to ensure that the whole building was made to the highest environmental standards. It achieved a BREEAM rating of Very Good, the highest achievable for a refurbishment. It also utilises passive air conditioning and 'ecoboard' made from agricultural residues through most spaces and reused 11,000 original bricks in the building (Derby Museums 2021).

During the design process, location had to be a major consideration in planning as the museum sits on the River Derwent. Newly installed flood measures such as polished concrete floors and cement particle board cladding, together with a new flood alert strategy, resulted in only one day being lost to the development programme despite two occasions of flooding during construction.

Adaptation and mitigation

There are museums across the UK that are already feeling the very real impacts of climate change. In recent years, major floods have continued to cause disruption to museum operations. Ironbridge Gorge in Shropshire which encompasses the River Severn is now facing floods at least once or twice a year which used to only occur every 10 to 20 years. The Ironbridge Gorge Museum Trust sites affected must take immediate action to remove objects and fittings, and, once flooding occurs, a lengthy and costly exercise in drying out and cleaning means the organisation must close to the public for an extended period of time, which has a knock-on effect on visitor income. Any future design will have to consider that the interior will have to be completely waterproof and demountable (Ralls 2022).

Leeds Museums and Galleries has similarly faced devastating floods, the worst being in 2015. A year on from the floods, Leeds Industrial Museum at Armley Mill hosted a community exhibition to commemorate them and asked the local community to join in by sharing memories and images (Leeds Museums and Galleries 2016). Now the recipient of an Arts Council grant from the Museum Estate and Development Fund (MEND, Arts Council England 2023), repairs to infrastructure and improvements in accessibility can be achieved to reopen the spaces affected (Sheridan 2022).

One organisation that provides support to the museum sector and others in managing issues of flooding and wider water management is the Canal River Trust which operates the National Waterways Museum. It has a team of hydrology specialists who provide advice and support on all water issues such as 'water resource use, water engineering and flood risk management' and 'embrace principles of efficiency, demand management and sustainability'. The Trust also has a team of ecologists who look after the canals (Canal and River Trust 2022).

The Trust provides a wide range of information online about how canals can fight climate change: providing active travel (walking, cycling), low carbon transport, reducing urban temperatures, creating wildlife havens and how canal water is used to heat and cool buildings. They are also looking at developing low-carbon boat energy and believe further investment in the infrastructures of their canals can help places around the country with flood management.

Tate has put in place methods to conserve water and reduce unnecessary wastage. Rainwater collected from the roof of Tate Modern is stored below ground in a 33,000-litre rainwater harvesting tank. Recycled rainwater is then pumped back into the building via submersible pumps located in the storage tanks. If the external tank should run dry during a drought, the internal tank also has a boosted cold water supply to keep the system operational. Within the building, the recycled rainwater is stored in a sectional tank, and a variable multistage booster pump set then provides regulated recycled water to each floor for flushing the toilets. Harvested rainwater and boosted cold water supplies are monitored and recorded on a monthly basis. An average of 226,000 litres of harvested rainwater is used annually to flush toilets at Tate Modern which equates to 37,666 flushes (NMDC 2021).

Considering that all museums in the UK are facing increased temperatures (which in summer 2022 temporarily closed some museums and galleries as the indoor spaces became uninhabitable during the 40-degree heatwave), all will need to make adaptations if they are to remain open to the public during such events.

Many industrial museums are seeking ways to ensure that they still provide an authentic experience to visitors and demonstrate historically accurate technologies in a more sustainable way. For museums with working machinery, particularly operating with coal and steam power, this provides a huge challenge.

London Museum of Water and Steam is developing an ambition for 'sustainable steam', looking at the possibilities for small, independent, industrial museums to become green. The museum, whose sites include 17 listed buildings and pumping stations, recognises the enormous challenge of running 200-year-old buildings that were not designed to be efficient. The site relies on a 92-year-old Lancaster boiler which has to be heated up for two days and cooled down for the same period of time. The museum acknowledges that there are some seemingly insurmountable problems but is determined to find innovative ways to move towards Net Zero as there's no other option to stay viable (Power 2022).

Using the principle that most good ideas are stolen, the museum hosted a hackathon with local engineering and building environment partners and young people, to look for green energy solutions. Solving problems in this way not only inspires the museum with eclectic ideas but also helps young engineers employed by the partners to understand how the work they do in the future can add to the betterment of society.

Workforce and travel

Carbon literacy training is now being widely rolled out across the sector, but this kind of training was given momentum by a partnership between Museum Development North West, Manchester Museum and the Carbon Literacy Trust, and their *Roots & Branches* programme which sought to dramatically scale up training in museums. The 'roots' are the museum, creating a nationally significant co-working hub of cultural environmental action to bring together environmentalists, museum staff, educators, third sector, students and others, and the 'branches' are the nationally aware and active museum sector. The project sought over the course of three years to certify as Carbon Literate 1,500 individuals who work and volunteer in museums and see 300 museums develop pledges to take action against climate change. Museums are seeking to address staff and business travel and to increasingly take transport, whether it be of artworks or people, into consideration in relation to their impact. Scope 3 emissions are one of the most complex areas for the museum sector to address. This category of emissions includes those from purchased goods and services, business travel, employee travel, waste disposal, transportation and distribution, investments and leased assets and franchises.

Whether and how museums take visitor travel into consideration in their environmental footprint is a complex issue. Not only it is difficult to obtain the relevant data, but many visitors use carbon-emitting vehicles to travel to museums. Though some museums have 'pivoted to digital' for an increasing proportion of their offer, the vast majority of museum activities remain in person, and so encouraging people to physically visit museums adds to carbon emissions. However, the positive impact museums have on people's lives and their own contribution to the climate crisis should be balanced with the impact of travelling to them. Museums may not always be able to determine the way visitors travel, but they can help promote sustainable travel options and encourage visitors to think about the journeys they make. Questions about whether and how museums include visitor travel in their footprints would therefore benefit from a sector-wide approach.

National Galleries Scotland (NGS) has been striving to become more sustainable, and transport has been a key part of the *Environmental Response Plan* (NGS 2021), particularly to address staff travel and making outreach work visibly more sustainable by using low-carbon transport methods. The Learning and Engagement Team created *Art in the Open,* a mobile art studio developed as part of the *Celebrating Scotland's Art* project. Support from the National Lottery Heritage Fund and Sustrans enabled artist Damian Callan to use an electric cargo bike to deliver free drop-in sessions all over the city and invite people to sketch the outdoors. Since then, the museum has acquired a cargo bike of its own and is making it available to staff for transporting equipment between sites as well as outreach work (NMDC 2021). The museum also has a small fleet of push and electric bikes, which adds to its green credentials.

Many museums have developed supported travel schemes to enable staff to purchase season tickets for train travel or to purchase bicycles. The Brunel Museum has taken the idea of sustainable staff travel and encouraging positive behaviours one step further by becoming the first museum to sign up to sustainable travel employee benefits scheme '*Climate Perks*' which allows employers to give paid 'journey days' which can be used to subsidise annual leave to take low-carbon transport modes to their holiday destinations. The aim of the scheme isn't just to get people flying less but to help them embrace 'slow travel' that sees them appreciate the journey as much as the destination (Brunel Museum 2022).

Environmental standards, collections storage, display and touring

As previously outlined, there are many elements which contribute to museums' difficulties in moving to more sustainable operations, but environmental standards for collections are the most 'museum specific' aspect. It has proven time and again to be one of the biggest stumbling blocks and requires sector-wide action.

The environmental standards adhered to by UK museums have developed over the last 40 years or so and have become a mainstay and expectation throughout the sector in managing collections for transportation, display and storage. They play a particularly key role in the lending and borrowing of works of art and objects dictating the conditions lenders expect to be maintained for insurance or indemnity agreements to facilitate loans and manage risk of lending work.

The systems used to maintain these conditions – HVAC (Heating, Ventilation, Air Conditioning) systems often referred to as 'climate control' – are energy intensive and costly. Their use has become standardised in museums across the world (though predominantly in the West) to maintain a regular window of temperature and relative humidity considered to be ideal for certain types of collections to ensure longevity and reduce the need for constant conservation.

This narrow window of 'ideal conditions' has expanded in the UK from suggested conditions originally used for artworks in the National Gallery's collection for 'preventative conservation' (Lambert 2014). Standards originally relating mostly to works on paper and for particularly vulnerable objects have become a more universally applied standard for all types of conditions, irrespective of whether a certain item or set of objects really requires them (see Southwick, this volume). When objects that potentially do need more support are often displayed alongside more hardy items, the blanket use of conditions becomes more convenient and, in some places, expected. But these conditions should never have been seen as 'one size fits all'.

The pursuit of these environmental standards has led to museums and galleries going to great lengths to provide at least one space and sometimes whole buildings in these conditions to be 'loan ready'. Arts Council England manages the UK Government Indemnity Scheme (GIS) which provides an alternative to commercial insurance for loans made accessible to the public in temporary exhibitions on long-term loans or loans for research and study. However, the environmental conditions required to qualify for the scheme act as a further driver to use HVAC to meet these standards in order for museums to be considered 'fit' to receive loans from national collections.

This of course has huge implications for cost, energy usage and sustainability for an organisation. There is also often a discrepancy in the conditions required of a borrower to be granted permission to lend, and the conditions objects are stored in by lenders themselves due to the constraints of their own buildings, heritage or funding situation.

However, change is afoot. After many attempts over the years to move beyond such strict parameters, there are concerted efforts across the UK and international sector to move to a more risk-based approach. Conservators, registrars, curators, directors and operations professionals are all moving to a different position whereby the shift in priorities for cultural organisations puts climate, people and sustainability on a more equal footing with preserving the collections. Networks of museum professionals across the UK and internationally are gathering evidence and working to change policy across the sector to relax standards. There is now even more impetus for change triggered by the energy crisis and the increase in bills which many museums simply cannot afford.

Conclusion

This chapter has looked at specific challenges and opportunities and areas where museums have taken action: in their role as civic institutions through public engagement, their capacity as research-based organisations and their estates and operations.

To move further forward, museums need sustainable funding to support this work. To address infrastructure needs to make buildings and operations more sustainable and to find and employ the right people or partners to deliver improvements or renovations, they need access to the right skills and investment for new roles or for upskilling existing staff. Putting together a sustainability plan is an endeavour in itself, and specific roles dedicated to sustainability are not something that all museums can afford, but it should be an ambition to ensure that someone has an overview of the needs of the organisation and can work with colleagues to drive change. All museum staff should have the opportunity to learn and engage in conversations about progress.

Support from existing funders is crucial to ensure that new funds are opened up that speak to the needs of the sector for adaptation, education and continued research and for existing funding sources to make grant recipients work harder towards environmental goals. Museums need to bring funders and supporters along on the journey and ensure that the push for sustainability comes from the top through boards and governance as much as throughout the whole organisation.

Museums in England in receipt of public funding via Arts Council England (ACE) as National Portfolio Organisations are required to submit sustainability plans as part of their funding agreement, and with 'Environmental Responsibility' as one of the four Investment Principles of ACE's 2020–2030 Strategy 'Let's Create' (ACE 2020), it can be expected that the scrutiny of how well funded organisations are performing against this goal will increase. However so far, there aren't many funders that are asking for explicit and demonstrable sustainability outcomes.

Similarly, the National Lottery Heritage Fund, a major funder of the sector, has increased focus on sustainability through its latest refreshed strategy and is committed to reducing the environmental impact and carbon footprint of the portfolio of projects and organisations it funds (National Lottery Heritage Fund 2023).

The theatre sector has developed the 'Theatre Green Book' (2021) which aims to support the transformation of organisations across three areas – productions, buildings and operations. The Green Book initiative has since been expanded with funding from Arts Council England and the Greater London Authority to a separate volume on Sustainable Buildings for a broader range of cultural organisations and includes a chapter on museums (Arts Green Book 2022). An accompanying 'Home Survey Tool' has also been provided by Buro Happold who developed the guidance to assess where an organisation stands in its sustainability journey and facilitates the creation of a sustainability plan. The guidance developed doesn't yet cover museum-specific collections issues but provides an outline of the considerations for adapting buildings and some of the simpler measures to introduce to improve performance.

The specific challenges of the buildings museums inhabit and the objects they care for within them do need to be acknowledged and addressed. Planning laws and heritage listing can seriously complicate necessary building works, as groups such as Grosvenor (2022) have outlined.

After many years of low investment in museum infrastructure, some capital is now available from central government via the MEND fund to help address the crumbling fabric of museum buildings. This money is currently targeted more towards emergency works rather than adaptation and improvement, but future rounds will hopefully provide the opportunity to build in sustainability. An impact of museums' heavy reliance on grant funding is that maintenance projects, unless paid for by core funding, tend to get neglected in favour of more attractive projects such as new buildings or public-facing programmes.

For museums it's rarely possible to simply turn everything off and lock up for a while. The challenges museums faced during the pandemic when unexpectedly asked to close down are magnified by the question of how to sustainably manage large esoteric collections in storage and display for the long term, while simultaneously maintaining a temperate climate for those who come to visit them, reducing their impact on planet and purse.

While sector-wide change is necessary such as adapting the conditions required for the approval of loans (e.g. via the ACE review of GIS), individual museums can act now. Many are already working with partners to widen the environmental parameters for loans and tours. Major lenders such as Tate are already committed to reducing the energy intensity that climate-controlled conditions create, and at an international level, the Bizot Group of museum directors has made a fresh commitment to adopting more flexible conditions after beginning this conversation over ten years ago (Tate 2021).

Making changes that were perhaps previously hampered by a fear of being the first to move, being liable or an outlier of widely accepted practices have now reached a point where change is inevitable and essential.

The challenges of reaching net zero and other environmental and biodiversity goals have come into much starker focus through the 'cost of living' crisis. Museums are very aware of the need to reach these goals, but in a sector that had not yet fully recovered from the impacts of the pandemic,[3] soaring energy prices and inflation have made the issue of paramount importance to their survival. The museum and other sectors can do so much, but the major push to green energy has to be driven by Government.

Despite the challenges, however, museums are engaged in massive changes and will always find ways to innovate. There are many groups and individuals working on efforts to instigate change such as the UK Registrars Group, ICON, the Museums Association, ICOM UK and Gallery Climate Coalition. Work is also underway at Arts Council England to review the conditions needed to qualify for the Government Indemnity Scheme.

NMDC is seeking to work with partners to bring together some of the more siloed action across the sector and to reach beyond museums to the visual arts and heritage sectors to address common problems including heritage buildings, exhibitions and displays, workforce skills, travel and Scope 3 emissions, ethical matters and funding. As these conversations and connections build, we will work towards action and commitments for museums within and beyond NMDC's membership to adopt.

Notes

1 NMDC's membership includes other non-museum-based national collections including RGB Kew, British Library, the British Film Institute, National Libraries of Scotland and the National Archives.

2 DCMS releases annual loan statistics via the Partnership Report; however, publication of the report for 2020/21 was delayed due to the pandemic. At the time of writing, the most recent figures reflecting a 'normal year' are from 2019/20; however, research loan statistics broadly mirrored the previous year while non-research loans for display were significantly reduced due to the restrictions of the pandemic. Figures relating to research loans are also only available for museums supported by government and covered by the report, but it is anticipated that many more museums lend their collections for these purposes but statistics aren't readily available.

3 The sector was in the process of 'bouncing back' when the energy crisis hit, with many museums seeing the return of around 70% of visitors in comparison to 2019 levels. However, the increase in energy costs will have far-reaching impacts not just on the operation of museum buildings and programmes, but also on the workforce and crucially on visitors who may now not continue to return in the numbers needed to support sector recovery. Some museums have seen energy bills increase by 400%, putting them in a precarious position. The future of Local Authority funding (the largest overall funding source for museums) and the situation regards energy bills will have wide-reaching impacts.

References

Arts Council England. 2020. *Let's Create*. https://www.artscouncil.org.uk/our-strategy-2020-2030

Arts Council England. 2023. *MEND (Museum Estate and Development Fund)*. https://www.artscouncil.org.uk/our-open-funds/cultural-investment-fund/museum-estate-and-development-fund-mend-round-3

Arts Green Book. 2022. https://artsgreenbook.com/

Arup. 2019. *You've Declared a Climate Emergency . . . What Next? Guidance for Local Authorities*. https://www.local.gov.uk/sites/default/files/documents/ARUP-Climate-Emergency-What-Next.pdf

British Council. 2019. *The Truth About Trust*. https://www.britishcouncil.org/research-policy-insight/insight-articles/truth-about-trust

British Standards Institute (BSI). 2012a. *Specification for Managing Environmental Conditions for Cultural Collections*. https://knowledge.bsigroup.com/products/specification-for-managing-environmental-conditions-for-cultural-collections/standard

British Standards Institute (BSI). 2012b. *Guide for the Storage and Exhibition of Archival Material*. https://knowledge.bsigroup.com/products/guide-for-the-storage-and-exhibition-of-archival-materials/standard

Brunel Museum. 2022. *Brunel Museum Becomes First Museum to Join Climate Perks Scheme*. https://thebrunelmuseum.com/brunel-museum-becomes-first-museum-to-join-climate-perks-scheme/#:~:text=The%20Brunel%20Museum%20has%20become,carbon%20transport%20options%20over%20flights

Canal and River Trust. 2022 *Nine Ways Canals Can Fight Climate Change*. https://canalrivertrust.org.uk/enjoy-the-waterways/canal-and-river-wildlife/nine-ways-canals-can-fight-climate-change

Collins, F. 2021. Museum Curators Among Top Five Most Trustworthy Profession, Poll Finds. *Museums Journal*. https://www.museumsassociation.org/museums-journal/news/2021/12/museum-curators-among-top-five-most-trustworthy-professions-poll-finds/

Crescent Arts. 2022. *Dear Plants. . ..* https://www.crescentarts.co.uk/whats_on/exhibition-dear-plants/

DCMS. 2021. *Museum Partnership Report, Sharing Collections 2019/20*. https://assets.publishing.service.gov.uk/government/uploads/system/uploads/attachment_data/file/1002386/2019_20_Sharing_Collections_v2.pdf

Derby Museums. 2021. *Annual Review 2020–21*. https://www.derbymuseums.org/news/derby-museums-annual-review-2020-2021

European Standards. 2018. *Conservation of Cultural Heritage*. https://www.en-standard.eu/bs-en-16893-2018-conservation-of-cultural-heritage-specifications-for-location-construction-and-modification-of-buildings-or-rooms-intended-for-the-storage-or-use-of-heritage-collections/

Grosvenor. 2022. *Heritage and Carbon: How Historic Buildings Can Help Tackle the Climate Crisis*. https://www.grosvenor.com/heritageandcarbon

Ipsos MORI. 2021. *Ipsos Veracity Index*. https://www.ipsos.com/en-uk/ipsos-veracity-index-trust-police-drops-second-year-row

Ipsos MORI Political Monitor. 2022. *Public Concern About Climate Change*. https://www.ipsos.com/en-uk/8-10-britons-concerned-about-climate-change-half-think-net-zero-target-should-be-brought-forward

Julie's Bicycle. 2022. *Arts Council England, Culture, Climate and Environmental Responsibility Annual Report 2021-22*. https://juliesbicycle.com/wp-content/uploads/2022/12/ACE389-Environment-Sustainability-2021-22-V7.pdf

Lambert, S. 2014. *The Early History of Preventative Conservation in Great Britain and the United States 1850–1950)*. https://journals.openedition.org/ceroart/3765

Leeds Museums and Galleries. 2016. *Appeal for Stories of the 2015 Leeds Floods*. https://museumsandgalleries.leeds.gov.uk/leeds-museums/armley-mills/appeal-for-stories-of-the-2015-leeds-floods/

Manchester Museum Research Collaborations. n.d. https://www.museum.manchester.ac.uk/collections/research/

Museum of Natural History Oxford. 2021. *Meat the Future*. https://oumnh.ox.ac.uk/learn-meat-the-future

National Galleries Scotland. 2021. *Environmental Response Plan*. https://www.nationalgalleries.org/art-and-artists/features/introducing-national-galleries-scotland-environmental-response-plan

National Lottery Heritage Fund. 2023. *Heritage 2033 – Our 10-Year Strategy*. https://www.heritagefund.org.uk/about/heritage-2033-strategy/download

National Museum Directors' Council (NMDC). 2021. *Green Museums: Tackling the Climate Crisis*. https://www.nationalmuseums.org.uk/what-we-do/contributing-sector/environmental-conditions/green-museums/

Natural History Museum. 2020. *A Planetary Emergency: Our Response: Strategy to 2031*. https://www.nhm.ac.uk/content/dam/nhmwww/about-us/our-vision/strategy-to-2031.pdf

Natural History Museum. 2022. *Econiario*. https://www.nhm.ac.uk/press-office/press-releases/giant-robotic-plant--econario--to-take-centre-stage-at-cop15-bio.html

Newell, J., Robin, L. and Wehner, K. (eds). 2017. *Curating the Future: Museums, Communities and Climate Change*. Abingdon: Routledge.

PK Porthcurno. 2022. *PK Porthcurno Win Double at National Cultural Enterprises Award*. https://pkporthcurno.com/pk-stories/pk-porthcurno-win-cultural-enterpises/

Power, L. 2022. Industrial, Independent, Small and . . . Green? *Presentation at the Museums Association Conference*. https://www.museumsassociation.org/event-highlights/conference-2022/ (paywall)

Ralls, N. 2022. Fire and Flood: Change Through Crisis. *Presentation at Association of Independent Museums Conference*.

Reading Museum. *Our Green Stories*. https://www.readingmuseum.org.uk/blog/our-green-stories

Royal Albert Memorial Museum (RAMM). 2022. *Seedscapes: Futureproofing Nature*. https://rammuseum.org.uk/news/seedscapes-future-proofing-nature/

Royal Botanic Gardens Kew. 2021a. *Our Manifesto for Change, 2021–2030*. https://www.kew.org/sites/default/files/2021-06/RBGK%20Sustainability%20Strategy_Final_June%202021_0.pdf

Royal Botanic Gardens Kew. 2021b. *Science Strategy 2021 – 2025*. https://www.kew.org/science/our-science/pubiications-and-reports/science-reports/kew-science-strategy

Royal Botanic Gardens Kew. 2021c. *Climate Positive by 2030*. https://www.kew.org/about-us/our-mission-and-priorities/sustainability

Science Museum Group. 2021a. *Towards Net Zero*. https://www.sciencemuseumgroup.org.uk/our-work/sustainability-approach/towards-net-zero/

Science Museum Group. 2021b. *Science Museum Group Announces Net Zero Target*. https://www.sciencemuseum.org.uk/about-us/press-office/science-museum-group-announces-net-zero-target-and-raft-new-sustainability

Sheridan, D. 2022. Leeds Industrial Museum to Receive Huge £653,000 Grant to Recover From Years of Flooding Damage. *Yorkshire Evening Post*. https://www.yorkshireeveningpost.co.uk/business/leeds-industrial-museum-to-receive-huge-ps653000-grant-to-recover-from-years-of-flooding-damage-3631223

Story:web. 2022. https://storyweb.info/

Tate. 2021. *Tate and Climate Change*. https://www.tate.org.uk/about-us/tate-and-climate-change

Theatre Green Book. 2021. https://theatregreenbook.com/
UKRI. n.d. Reef Refugia (NHM). https://gtr.ukri.org/projects?ref=NE%2FR011044%2F1
University of Cambridge. 2022. *Butterflies Through Time*. https://www.museum.zoo.cam.ac.uk/butterflies/butterflies-through-time
V&A. 2021. *Sustainability Plan 2021–24*. https://www.vam.ac.uk/info/sustainability

6

MUSEUMS TACKLING CLIMATE CHANGE

A Zimbabwean context

Simbarashe Shadreck Chitima

Introduction

The United Nations Intergovernmental Panel on Climate Change (2007b), through a special report published in 2018 and 2022, indicates that due to climate change the world is likely to experience a temperature rise of 1.5°C by 2040. The world is under threat from climate change which has destroyed the human and biota ecosystems. Scientists around the globe indicate that the majority of problems society faces are caused by climate change. Climate change refers to changes in temperatures, rainfall and wind conditions which are altered mainly by variations in radiation, volcano activity, plate tectonics, global warming, changes in the gas composition of the atmosphere and shifting and wearing away of land surfaces and oceans (Walmsley 2010). It has also been scientifically proven that climate change is mainly caused by human action such as deforestation, gaseous or chemical leaks and burning of fossil fuels (coal, gas and oil) leading to emissions of gases such as carbon dioxide and methane into the atmosphere. This amplifies a greenhouse effect leading to temperature warming, heat waves, snow and ice cover and cold snaps among other effects (IPCC 2007a, 2007b). African countries are prone to many effects of climate change because they are mismanaged, ill-equipped, under-resourced and ignorant as many politicians concentrate on power politics (Maipose 2000). Zimbabwe has experienced several climate-change-induced disasters such as droughts, floods and cyclones. For example Zimbabwe experienced droughts in 1982 and 1992, an earthquake in 2007, cyclones Dineo (2017) and Idai (2019). Cyclone Idai was devastating as it caused substantial flooding, resulting in numerous human deaths and livestock, damaged infrastructure property and affected the operations of heritage institutions (Ndiweni and Musarurwa 2014).

DOI: 10.4324/9781003347606-7

The challenges posed by climate breakdown can prompt a pessimistic outlook which cannot imagine trouble-free future, causing some to vow not to have children if global leaders, politicians and public institutions do not act accordingly (Verline 2021). Environmental consciousness and activism have existed since the 1960s but have increased to new heights led by young people in the 21st century. The environmental youth movement that takes prominence from the likes of Greta Thunberg from Sweden has staged 'Fridays for Future' protests, absconding from school for Fridays of the school term to highlight the needs for governments and institutions of power to take action. Greta Thunberg and others such as Vanessa Nakate from Uganda have inspired many young people from different countries in climate strikes around the world. These youths understand that if action is not taken now they will be left with an uninhabitable world of natural disasters, unclean air quality, pandemics, unclean water sources, poor soils and degraded farming land.

The effects of climate change will not spare any organisation, including museums, and it is imperative that action be taken to respond to the threats to collections, sites and monuments posed by the climate crisis. This study examines the role of national art galleries and museums in Zimbabwe in mitigating the effects of climate change.

Environmental conventions and heritage management

Consciousness to protect the environment is clearly advocated for in the 1970s particularly during the development of ecomuseums. The ecomuseum concept is synonymous with Georges-Henri Riviere and Hugues De Varine, who encouraged the linkage of heritage protection with the environment (De Varine 1985). The 1972 World Heritage Convention provides the ground for world leaders to invest in the protection of heritage property and the environment. Other conventions and fora that emphasised for the protection of the environment are the 1985 Vienna Convention for the Protection of the Ozone Layer, the 1997 Earth Summit and 1997 Kyoto Conference, the 2003 African Convention of Nature and Natural Resources, the 2015 United Nations Sustainable Development Summit and the 2016 United Nations Paris Agreement, which speak the need for protecting the environment and heritage. The Sendai Framework for Disaster Risk Reduction (2015–2030) of 2015 also aims to promote disaster risk management. These environmental and heritage conventions acknowledge the devastating effects of climate change but place human beings at the centre of sustainable development.

Climate change is a major threat to the world and its heritage; however, a sustainable future can be achieved through embracing a green ethos to prevent catastrophic deterioration and destruction of heritage property (Brophy and Wylie 2008). Green measures have been encouraged as these are the point of contact through which museums can contribute to climate change mitigation (Chitima 2015).

The green concept

Green museums are those that incorporate concepts of sustainability into their operations, social design and programming (Brophy and Wylie 2008). According to these authors, greening in museums is an approach of tackling climate change through the deployment of eco-conscious and environmentally benign strategies. A green ethos is a culture (mindset, behaviours, actions and measures) that prevents deterioration and destruction of cultural property from the effects of climate change (ibid.). The aim of employing a green ethos assists in reducing the impacts of our operations and conduct towards the destruction of the environment. However, as museums implement green measures, they stand to benefit by reducing costs, operating effectively and improving the social and political standing of heritage institutions. This has an ultimate positive result which leads to sustainable heritage institutions that are able to manage heritage property –collections, sites and monuments while reducing their carbon footprint and climate change consequences. A green ethos is grounded on the Intergenerational Equity (IGE) philosophy where heritage institutions have the responsibility to effectively manage heritage property so that current generations access and use the heritage in a sustainable manner which will not deny future generations realising the same benefits (Weiss 2013). This study employs the IGE philosophy as its conceptual framework.

Green strategies that museums can employ include water, waste and energy management, maintaining green buildings, green procurement, green exhibit design and green-themed exhibitions (Byers 2008). Dedeoglu (2020) indicates that the world is now facing droughts, and the majority of water sources are becoming polluted. Rubbish, waste and chemicals used in carrying out remedial conservation are a hazard to the environment if not properly managed and may lead to the pollution of water sources. Therefore, there is a need for water management that may include rain water harvesting, reusing black water (used water) or using underground water extracted via boreholes using solar power.

Museums use a large amount of energy, almost all generated via fossil fuels, in heating, lighting, air conditioning and electrical services aimed at providing suitable environments for collections and visitors in galleries. Future energy provision thus must include solar, wind, bio, hydroelectric and geothermal energy (Horse 2018). Energy management can also be tackled by maintaining green buildings that allow natural light to filter through providing cheap energy for museums (Dedeoglu 2020).

Another green challenge for museums is how to avoid waste generated from museum activities and operations leading to environmental pollution. For example some museums have food cafés, and a lot of waste such as perishable material and left overs are generated. As such, instead of throwing away this material contributing to environmental pollution, the same material can be composted (Byers 2008). Waste material found in museum food cafés for example can be composted, and the fertiliser created can be used in the museum landscape. Byers (*ibid.*) indicates that waste management should be undergirded by the cradle to the cradle theory (reuse or recycling of material).

Brophy and Wylie (2008) argue that museums can also contribute to disaster reduction and climate change education through the use of green-themed exhibitions and displays, publications, study packs, talks, educational and public programming with the aim to share climate change information (Byers 2008; Dedeoglu 2020). Information and data are highly valued resources and, if made available, will assist in policy framework and behaviour change among local communities. Thus, museums can fill this gap and facilitate research and data availability and its accessibility. Museums can contribute to environmental protection by making use of materials that have low volatile organic compounds in exhibition design (Dedeoglu 2020).

Significantly, however, the strategies and ethos presented earlier are the ones established in Western museums and communicated in Western publications. They strategies may not automatically apply to all heritage institutions in different contexts around the globe due to differences in museum location, resources availability, existence of political will, staff motivation, knowledge and training as well as policy frameworks. To that end, it is interesting to know how museums in Africa's developing countries are contributing to climate change mitigation.

Methodology

The study was undertaken from 2016 to 2019. The study employed qualitative research and multiple case studies that included the National Art Galleries in Harare and Bulawayo, the Natural History Museum, the Batonga Community Museum, the Zimbabwe Museum of Human Sciences and the Great Zimbabwe World Heritage Site. The study deployed observations, face-to-face and social media interviews as well as green audits as research instruments. The green audit specifically focused on the nature of museum buildings; conservation measures; merchandising; waste, water, energy management; green policy; exhibition design and educational programming. The study population included a total of 34 participants drawn from four art gallery and museum directors, six curators, five exhibition or display designers, six education officers, nine tour guides or attendants and four officers from the Ministry of Environment, Water and Climate (MEWC). Although art galleries and museum administrators were not willing to divulge sensitive information such as how much they were spending on water, energy and waste management, information on several green facets was acquired from curators, tour guides or attendants, conservation officers and exhibition or display designers.

Results

This section is thematically arranged into two categories dealing with the green measures and ethos museums are employing and the barriers they face.

Museum green measures

There are several green initiatives that national art galleries and museums in Zimbabwe are employing. These include staging green-themed exhibitions, research and symposia, green conservation measures and water and energy management. The most common activity is staging green themed exhibitions. The National Art Galleries in Harare and Bulawayo have staged several and annual green exhibitions in conjunction with the Environmental Authority management on green topics since 2016. National Art Galleries of Zimbabwe in Harare (NAGZH) and National Art Gallery of Zimbabwe in Bulawayo (NAGZB) as well as the Natural History Museum (NHM) have staged several exhibitions on the protection of the environment from 2016 to 2019. NAGZH staged an annual exhibition titled 'Green Shoots' to educate visitors from early childhood pupils up to 'A' level students about climate change. Further, in 2020, NAGZH developed an exhibition titled 'After Shock: Life After Cyclone Idai' mainly raising awareness on the impacts of climate change. The NAGZB has also staged a national and regional exhibition in 2018 titled 'Show Your Care' that communicates about environmental protection. Both the two national art galleries have staged a vigorous campaign to encourage artists to recycle found objects to make art.

The NHM has on several occasions hosted climate change exhibitions and included questions on climate change in the quiz programme for primary school students. The NHM has also hosted several symposia where researchers across the country shared information on climate change. The museum has also provided educational sessions to primary and secondary students on climate change, the use of green strategies to protect the environment and heritage found in Zimbabwe. Two environmental experts at Midlands State University indicated that climate change education is important when taught by museums because they have materials which they can use to facilitate experiential learning. These educational initiatives were aimed at encouraging members of society to lead a sustainable lifestyle.

National art galleries in Zimbabwe differ from the national museums in their approach to lighting. While the museums rely on energy provided by the Zimbabwe Electricity Supply Authority and limit their energy-saving measures to switching off the lighting when the galleries are not in use, the architecture of the art galleries as a traditional approach to illumination allows natural light to filter through the galleries.

The Batonga Community Museum (BCM) has invested in using indigenous knowledge in the conservation of collections. The BCM has had problems with infestations of bats and insects that led to bio-deterioration. Fumigation has caused several conservation problems in relation to organic collections, so other methods were examined to lessen deterioration of collections. Traditionally, local communities have used a variety of ways to tackle these issues. In 2015, the BCM made use of chilli and burnt charcoal to force bats and insects to migrate from the museum to other areas, and therefore indigenous knowledge did not contribute to the harm of collections and humans.

The nature of museum building contributes to the effectiveness of collections care. Poorly insulated buildings are always affected by temperature, light and relative humidity as well as pollution. The BCM is able to effectively control temperature and relative humidity because it has a thatched roof. The research showed that it was only the BCM that involved local communities in green conservation rather than make use of expensive conservation measures that were detrimental to collections.

In relation to natural disasters or more extreme weather events, cultural heritage sites and museums are using a strategy of adaptation. This involved acknowledging that natural disasters threaten heritage sites hence fire zones have been created around cultural heritage sites while staff members have been trained on rescue strategies. Other climate adaptation strategies employed by museums in Zimbabwe include educating and co-managing cultural heritage sites with local communities so that they are well preserved. At Great Zimbabwe, Khami and Matopos World Heritage sites, several projects have been undertaken to clear grass and the invasive lantana camara around the site perimeter to limit the effects of veld fires. Lantana camara was seen to have adverse effects on the environment where it forms dense thickets killing or excluding native plant species and dominating the understory shrub. Lantana camara is also known for increasing fire risk to heritage sites especially during the dry season. Education has gone a long way to inform local communities on the importance of preserving flora and wetlands near cultural heritage sites. Further, other strategies deployed by museums have been to establish and formulate communication plans for disaster mitigation and management with various stakeholders such as the fire brigade, emergency services and the police or security services. Museum security and tour guide conduct physical daily inspections and provide physical security.

Barriers

It was gathered that all national museums and art galleries in Zimbabwe operate without formal green policies and are pursuing individual measures in the mitigation of climate change effects. This was a barrier to the implementation of green measures across the heritage sector. The study further gathered that museum staff members were not aware of the concept of green or sustainable museums. When asked about the green initiatives the NMH was employing, tour guides indicated that they were not aware of the issue. The education officer at the NHM reported that communication of measures mitigating against the impact of climate change has highly been individualistic in museums and that the majority of staff members were not aware of them. They further stated that policy frameworks were needed to inform and educate museum staff members of the steps to take in the effort to develop and implement an integrated approach to climate change. The study also gathered that museum and art gallery staff members had little motivation to seriously engage in the discourse of mitigating climate change impacts because there was a lack of political will from senior management. One curator indicated that senior management did not budget for the implementation of green measures at his museum, and hence the middle and lower staff members professed ignorance of the issues.

Closely tied to the issue of lack of budgets, it was also revealed by museum curators that national museums in Zimbabwe do not have adequate infrastructure in terms of buildings, storerooms and advanced conservation equipment to carry out their work. Three museum directors indicated that heritage institutions in Zimbabwe were suffering due to the fact that they were not prioritised by the state when it came to funding, and hence this also affected even the brightest ideas and projects they wanted to implement. Therefore, the major conclusion that derived from the interviews with museum and art gallery directors was that effective planning and response to climate change should begin at the government level, then cascading to government institutions and on to the local communities. For example if museums were to fully convert their buildings to solar power, there was a need for government to provide more funding and to place heritage institutions in national energy grids where electricity is not rationed. The NHM faces several conservation challenges due to the intermittent power supply it faces which has contributed to the deterioration of frozen natural history material such as the scorpion and frog collections. The use of generators to power the museum has also contributed to pollution. The live snake collection at the NHM requires lighting to keep them warm, but due to power cuts the museum has been struggling to manage the collections. Power cuts experienced at the NHM have affected collections conservation and visitorship.

Therefore, a lack of political will, resources, knowledge and training were cited as contributing to museums and art galleries failing to effectively embrace the discourse of climate change mitigation and disaster risk management. The lack of policy frameworks and separate ministerial placement of national museums and art galleries has led to unstructured as well as uncoordinated efforts towards climate change response. Another key point was that museums were thought as engaging in climate change preparedness and response without the involvement of local indigenous cultures. As noted, it was only the Batonga Community Museum that partly involved local communities in dealing with bats and insects that had posed a conservation challenge to collections. The other art galleries and museums have not yet democratised their curatorial spaces to fully involve local communities in environmental protection and mitigate against the effects of climate change.

Museum Concentric Climate Change Response Framework (MCCCRF)

A framework is a basic structure underlying a concept intended to support the building of ideas and theories. This study has developed a Museum Concentric Climate Change Response Framework which establishes the major factors that influence the effectiveness of museum response behaviour to climate change. The diagram in this Framework is indicated as being concentric in the sense that even after implementing green measures, there is need to review efforts, revisit the issues and conduct risk and green audits. This framework is established after gathering information from environmental management experts and museum staff members in Zimbabwe for four years as well as observing museum operational practices. It establishes the major components that museums need to deploy in order to effectively mitigate and respond to climate change. The diagram in Figure 6.1 illustrates the MCCCRF.

Heritage institutions in Zimbabwe are taking a back seat stance due to lack of political will by government and senior museum management. The lack of political will is evident by the lack of policy frameworks, funding and co-ordinated effort towards reducing the effects of climate change. Any successful measure to mitigate the effects of climate change requires collaboration of government and museums. The government is an important player when it comes to environmental and cultural heritage protection. Many African countries including Zimbabwe however concentrate on politicking where they blame developed countries for largely destroying the environment and atmosphere through pollution from industries. There are few industries in Africa that pollute the same as developed countries, hence many governments in Africa seem to believe that climate change is a foreign phenomenon. National policy frameworks that seek to protect the environment in Zimbabwe are selectively implemented. For example Chinese companies have a large stake in the mining and industrial business, and these are known for wanton land degradation and pollution (Maipose 2000). The Chinese are the chief patron of the Zimbabwean government, and it has become very challenging for the state to police Chinese companies for contributing to the destruction of the environment

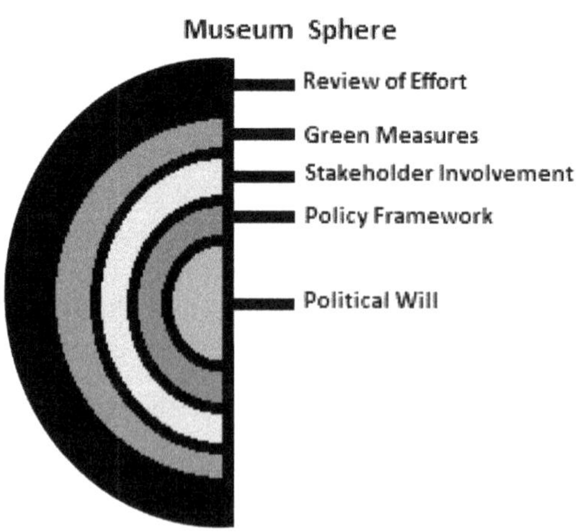

FIGURE 6.1 The museum concentric climate change response framework

through pollution and land degradation evidenced by their mining activities. It is such situations where there are policy implementation inconsistencies that weaken the state's effort towards demonstrating to other government and non-government entities the seriousness of mitigating against climate change. When the government is compromised, it has become difficult then to ensure that all other institutions comply and align with national policies. This, in turn, leads to the lack of political will by government to effectively implement policy that seeks to protect the environment. There is need for the government to take a stance and request every institution and person to act accordingly in mitigating against climate change and destruction of the environment (Maipose 2000). There is need also for government to provide monetary resources, enact laws and formulate implementation strategies that will obligate every institution including museums to take action against engaging in operations and lifestyles that are not green and sustainable.

There is also lack of political will by museums' senior management, which sees an absence of green policies and green teams to lead in the implementation of coordinated efforts. The lack of political will at museum level has led in turn to the lack of prioritised budgeting and resourcing so that green measures are effectively implemented.

The MCCCRF also indicates the fact that it requires collaboration of museums and different stakeholders including local communities, government and government agencies dealing with environmental protection and disaster risk management. Indigenous knowledge is a valuable knowledge that has been found not to be harmful to museum collections as compared to expensive remedial and conservation treatments (Chitima and Ndlovu 2018). Indigenous knowledge is usually shunned as unauthentic because the world has so much been used to using Western knowledge systems (Horse 2018). Most often, information on climate change is disseminated by scientists, yet there are other reliable sources of information, expertise and experience such as involving local communities in the management of cultural property. Indigenous knowledge from local communities is an effective body of knowledge that museums can use in the conservation of collections or disaster preparedness. There is great opportunity for cultural institutions in acknowledging and promoting indigenous knowledge systems and adaptation strategies. Mukanganise (2011) indicates that indigenous cultures have rich information that contributes to detecting early signs of natural disasters. For example local communities in Mbire, Zimbabwe, have shown that when there is excessive fruiting by wild trees or watermelons, it indicates that there is a going to be a drought in the same year (ibid.). When birds that usually build their nests on low branches along the river start building their nests higher up, it indicates that there will be flooding in the area (ibid.). This knowledge generated locally has been proven or validated for so many years and hence can be vital when museums involve local communities in climate change mitigation. Such indigenous knowledge can be harnessed to develop disaster preparedness initiatives and green measures to be used in effective management of cultural property.

Climate change management can only happen effectively when there is collaboration of government, museums and other government agencies. The society at large is an important stakeholder and player in issues that affect their lives. Effects of climate change have the potential to destroy human lives, flora and fauna as well as domestic animals which many African people view as wealth and food sources. When the society is not involved in debates, talks or discussions about climate change, environmental and heritage protection, it may continue to lead unsustainable lifestyles like causing deforestation and pollution among other actions and behaviours.

Museums are in a position to reduce the effects of climate change when they also include local communities and adopt indigenous knowledge systems. The BCM was able to avoid the expensive and harmful use of chemicals when the museum faced challenges with insects and bats. Instead by acknowledging the utility of indigenous knowledge, heritage institutions can benefit from locally generated knowledges that are green. However, as museums engage with different stakeholders, there is need for training of players including museum staff so that they are equipped and empowered with the same knowledge and skills for effective implementation. Training avoids contestations during implementation and enhances coordination of efforts.

National art galleries and museums have used exhibitions as a green measure to facilitate the teaching of climate change topics. This has been a success because museums are educational learning settings for the young generation of children and students. Museums can be used to have agency in climate change because of their communicative, affective and social qualities (Cameron et al. 2015). Museums are regarded as trustworthy information centres, and many governments view them as communicative vehicles for climate science. Selected national art galleries and museums in Zimbabwe have contributed to information dissemination on climate change. Many societies engage in unsustainable behaviours and actions because they are less informed on climate change. Thus, the major green measure to communicate climate change education in Zimbabwe has been that of using green-themed exhibitions, hosting talks and symposia. Suarez and Tsutsui (2004) indicate that museums are trusted institutions that can provide information to societies they serve. There are several opportunities which museums can use to learn and promote research about climate change and pandemics. Museum collections can be used as reference sources that can be used in the study of pandemics, biological terrorism, public health and climate change.

Research on biodiversity and loss, biological invasions, pathogens, vectors of diseases and environmental contaminants can be done using museum collections. Museums can take the lead role in research about climate change, how society has adapted to it, and this information can inspire future research and lead to knowledge production. Museum staff members have the opportunity to produce publications on climate change and environmental protection which can be accessible to members of the public. Collections at the Smithsonian Museums have been used to study HIV, while at the Bishop Museum in Hawaii, mosquito collections have been used to study malaria and West Nile Virus (Suarez and Tsutsui 2004). Museum collections can be used to track and study pandemics, habitat loss, virus strains and impacts of climate change on different sectors of life. Brophy and Wylie (2008) indicates that another strategy that cultural institutions can implement is to use sustainable materials in exhibition making, such as the use of wheat board and bamboo material in exhibition construction because these emit low volatile organic compounds. Boards with adhesives and unseasoned woods that make up the largest percentage of shelf-making material are detrimental and lead to the deterioration of collections as they produce high relative humidity and volatile organic compounds.

Museums can also be change catalysts by providing space for debates and symposia. The NHM continues to host symposia on climate change, and this strategy generates information through research presentation and debates. Museums and art galleries in Zimbabwe are well positioned to be information disseminators through online platforms and on television. Due to technological advances and the social distance regulations in many countries, information can easily be spread through social media and online platforms. So far, national art galleries and museums in Zimbabwe have not embraced the use of online facilities to give education on climate change. Cameron et al. (2015) indicate that museums can also communicate climate change information through publications and educational programming for school students. The curriculum in Zimbabwe for early childhood, primary and secondary students has topics on climate change and environmental protection. This aims to inculcate in younger generations environmental consciousness and a sustainable future. National art galleries and museums in Zimbabwe provide outreach programmes, but these aim to propagate in primary and secondary students the type and nature of heritages found in Zimbabwe. It is this foundation which art galleries and museums may also build to include topics related to climate change and environmental protection. Although the number of questions relating to environmental protection in the school's quiz programme is small, but fusing curriculum-related questions that speak about climate change is an important step.

As noted earlier, art galleries in Zimbabwe encourage artists to recycle materials in their productions. New types of art presentation and production styles can be encouraged utilising new technologies. Cameron et al. (2015) posit that art can be used as a medium to converse about climate change information. For example an art project, the 'Creative Climate', launched by the BBC World Service and the Open University, features diary entries where personalities can record their trajectories and personal experiences with environmental change (ibid.).

Planting within landscapes is another green initiative which art galleries and museums can use to promote quality working environments. Having a culture of planting plants in the museum environment and outside landscape contributes to the generation of clean air (Byers 2008). Indigenous fruit trees can be planted within the museum landscape with the benefits that their fruits can be consumed by staff members and that they help block dangerous winds that may destroy infrastructure.

Following the implementation of any green measure, there should be a review to ascertain their success (Chitima 2015). Over the last 50 years, the world has been battling with effects of climate change which have taken a toll on cultural heritage management. There is need for everyone and every institution to employ green measures in mitigating against the effects of climate change. Museums have the capacity to engage and mitigate the risks associated with climate change at micro and macro levels. At micro level, museums can deploy green initiatives that directly benefit their operations without increasing their carbon footprint. At macro level, museums have the capacity to communicate climate change education to society. There is room for heritage institutions in Zimbabwe to do more to reduce their carbon footprint by having clear institutional frameworks to structure their work in a coordinated way.

References

Brophy, S.S. and Wylie, E. 2008. *The Green Museum: A Primer on Environmental Practice*. Plymouth: Altamira Press.

Byers, R. 2008. *Green Museums and Green Exhibits: Communicating Sustainability through Content and Design* (Unpublished Masters of Science in Arts Administration Dissertation, University of Oregon).

Cameron, F.R., Hodge, B. and Salazar, J.F. 2015. Climate Change Engagement a Manifesto for Museums and Science Centers. In Cameron, F.R. and Neilson, B. (eds) *Climate Change and Museum Futures*. New York: Taylor and Francis, 248–275.

Chitima, S.S. 2015. Developing Sustainable Museums Through 'Greening': A Case Study of the Zimbabwe Military Museum. In Mawere, M., Chiwaura, H. and Thondhlana, T.P. (eds) *African Museums in the Making: Reflections on the Politics of Material and Public Culture in Zimbabwe*. Bamenda: Langaa Research and Publishing, 223–246.

Chitima, S.S. and Ndlovu, I. 2018. Incorporating Indigenous Knowledge in the Control of Wahlberg Bats at BaTonga Community Museum as a Collections Preservation Measure. In Ngulube, P. (eds) *Handbook of Research on Heritage Management and Preservation*. Hershey, PA: IGI Global, 396–407.

De Varine, H. 1985. Word and Beyond. *Museum*, 37(4): 185.

Dedeoglu, D. 2020. *Green Museums: An Introduction and a Possible Implementation in Ankara* (MA thesis submitted to Ihsan Dogramaci Bilkent University, Turkey).

Horse, J.N. 2018. *Green Museums Waking Up the World: Indigenous and Mainstream Approaches to Exploring Sustainability* (PhD submitted to Montana State University, USA).

Intergovernmental Panel on Climate Change (IPCC). 2007a. *Climate Change 2007: The Physical Science Basis*. Contribution of Working Group I to the Fourth Assessment Report of the IPCC. Geneva: IPCC.

Intergovernmental Panel on Climate Change (IPCC). 2007b. *Climate Change 2007: Mitigation of Climate Change*. Contribution of Working Group 111 to the Fourth Assessment Report of the IPCC. Geneva: IPCC.

Maipose, G.S. 2000. Aid Abuse and Mismanagement in Africa: Problems of Accountability, Transparency and Ethical Leadership. In Hope, K.R. and Chikulo, B.C. (eds)

Corruption and Development in Africa. London: Palgrave Macmillan, 87– 103. https://doi.org/10.1057/9780333982440_5.

Mukanganise, R. 2011. *Disaster Preparedness at Community Level in Zimbabwe: The Case of Chirumanzu and Mbire* (MA dissertation Submitted to the Women's University, Zimbabwe).

Ndiweni, N.J. and Musarurwa, C. 2014. Natural Hazards as Disasters: Mitigation and Challenges in Southern Zimbabwe. *Journal of Emerging Trends in Education Research and Policy*, 5(1): 14–19.

Suarez, A.V. and Tsutsui, N.D. 2004. The Value of Museum Collections for Research and Society. *BioScience*, 54(1): 66–74.

Verline, B. 2021. *Learning to Live With Climate Change: From Anxiety to Transformation*. London: Routledge.

Walmsley, D. 2010. *Climate Change and Its Effects on Humans. State of the Gulf of Maine Report*. http://www.gulfofmaine.org/state-of-the-gulf/docs/climate-change-and-its-effects-on-humans.pdf

Weiss, B. 2013. *Intergenerational Equity*. https://opil.ouplaw.com/view/10.1093/law:epil/9780199231690/law-9780199231690-e1421 (Accessed 4 February 2020).

7

THE GLOBAL SOUTH EMERGES

How cultural institutions in South America are using storytelling to call audiences to action in tackling climate change

Eduardo Carvalho

Introduction

From the rock paintings that adorn the walls of caves across the world to the most technological experience currently available, like the metaverse, storytelling has been an essential part of human life to share messages that provoke reflections and change. This capacity to influence must be used if we hope to change the society in which we currently coexist.

We are facing a climate emergency due to rising global temperatures, caused by the emissions of greenhouse gases into the atmosphere. Our generation is facing one of its greatest challenges.

Considering this reality, museums across the globe have taken action to scale the narrative on the urgent need to deal with the climate crisis. Even though many institutions have declared a climate emergency and have prepared bold plans for energy transition or for restructuring their collections, the need to create meaningful engagement with audiences has grown more urgent, to facilitate communication and the language that contributes to greater awareness and to make content more identifiable among audiences.

Humanizing messages, with empathy as raw material, is necessary to create a context for facts and policies that influence all aspects of life. This may be effective in the so-called call to action, making it part of the solution, without blaming an individual for the current dilemmas.

A punch from reality

The IPCC (Intergovernmental Panel on Climate Change) released its most recent report (IPCC 2021), providing indisputable evidence that the increases in

DOI: 10.4324/9781003347606-8

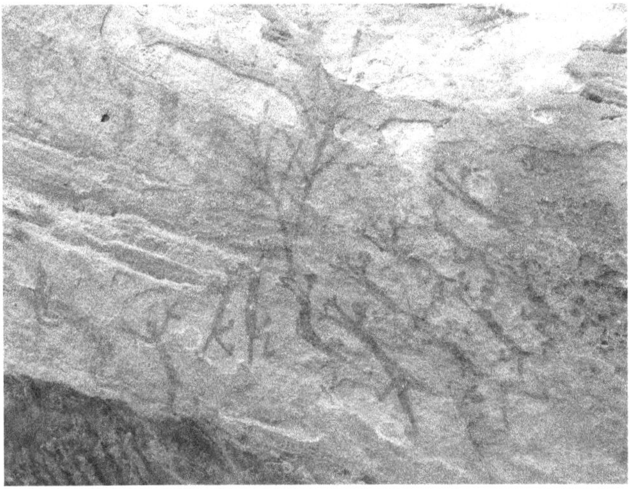

FIGURE 7.1 Rock shelter in the Serra da Capivara National Park, Brazil, is decorated with cave paintings, some more than 25,000 years old. Photo credit: author

greenhouse gas concentrations recorded since 1750 are caused by human activity. This has led to a 1.1°C increase in the Earth's surface temperature from 1850 to 2019.

From the perspective of the Global South, which describes developing nations in Africa, Latin America and the Caribbean, Asia and Oceania (whether large economies, like China and Brazil, or small economies), the effects of climate change are increasingly present and severe and are already being reflected in public opinion.

A study conducted by Ipsos (2022) for the World Economic Forum found that, on average, across 34 countries, more than half of all adults surveyed (56%) said climate change has already had a severe effect in the area where they live. The highest concern came from the Global South (Mexico, with 75%). Regarding the chance of being displaced from home as a result of climate change at some point in the next 25 years, India presented the highest concern (65%). Other countries, like Brazil (49%) and China (41%), also presented high numbers when compared to developed nations like Sweden (17%) or Great Britain (30%).

Most of the highest levels of concern are found among populations that live in countries close to tropical forests, home to important ecosystems containing almost half of all terrestrial biodiversity, which is being impacted by human action. One example is the Amazon, the world's largest tropical forest, stretching over eight countries in South America. It is currently suffering from the effects of deforestation, the invasion of public areas and increased criminal activities in territories belonging to indigenous peoples, "quilombolas" (Afro-Brazilian maroon villages), and riparian communities, not to mention climate change, accelerated by the global increase in greenhouse gas emissions (Abramovay 2019).

Climate change, accelerated by human actions, can forever transform this and other forests and their biodiversity by causing increased temperatures, decreased productivity of certain crops, reduced rainfall and more fires affecting natural vegetation. The geological phenomenon underway is the result of a society that has lost respect for nature's boundaries, having disconnected from the environment that provides the resources that sustain life. We occupy everything around us, connected between us, though we remain somewhat ignorant of our relationship with other species that we share this planet with.

Transforming data into feelings

To develop storytelling, that is an unforgettable narrative, you need to listen to what the story is telling you (Buster 2013) and ask yourself what the truth is inside this story to be told. Furthermore, according to the author, all stories that we tell have a journey that seems to have an uncertain end, though one filled with hope and courage. The author concludes that this is how we connect with our humanity and become better people. The way we tell a story can make all the difference in anything that you do (Buster 2013).

Becoming part of the experience is one of the leading calls behind the concept called the "Experience Economy", coined by Joseph Pine II and James Gilmore in 1998. They explained that the development of experiences is no longer exclusive to the theme park and theatre industries, traditionally geared towards this concept, and is now part of the goods and services sector, which offers products gauged not by what they are actually worth, but instead by the intangible, thereby creating a new source of value.

Pine and Gilmore argue that employing new technologies encourages the creation of new genres of experience, which must be developed with excellent design, a marketing campaign and correct operation, traits fundamental to a goods and services consumer, who now also seeks this in experiences.

One of the essential points in developing an experiential project is the need to get out of the office and speak to potential users, seeing things from their point of view and attempting to act as they would (Lewrick et al. 2018). This practice will aid the development of an empathy map in which the developer (or Design Thinker) will have to think, feel, listen, speak, act and see like the potential user, taking into consideration the pros and cons throughout the process:

> Especially when we think about learning more about the products or situation, we approach the situation as a novice – curious and without any prior knowledge. Consciously, and with all our senses, we go through the same user experience . . . it is useful to define hypotheses in the team, test them with a potential user or use existing data, then confirm, discard or adapt them.
>
> *(Lewrick et al. 2018: 28)*

One of the important points for undertaking this project is the application of the concept of "edutainment", a neologism combining education and entertainment. According to Pine and Gilmore:

> As is the case with entertainment experiences, educational experiences lead users (or students, if you prefer), to absorb the events that develop before them, on the contrary to entertainment, the educational experience implies an active participation by the subject. To inform how to correspond to a person and increase their knowledge and/or skills, educational events must actively involve their minds (if an intellectual education) and/or their body – if involving physical training.
>
> *(Pine and Gilmore 1998: 70)*

Upon reviewing the "Experience Economy" concept, Pine and Gilmore reiterated the importance of experiences provoking transformation in the user. The authors maintain the definition that by customizing a product – creating a difference between it and commodities – goods and services become unique experiences to users. Even en masse, they are perceived on an individual level (Pine and Gilmore 2013).

However, following criticism, both agreed that experiences also had to transform the user. This not only reduces the risk of the experience falling along the wayside, it also provokes a sense of "wanting to change" in the user. The authors affirm:

> The experience economy is an underlying long-term change in the actual framework of advanced economies – and the forces of creative destruction take time. New forms of economic production don't just appear automatically. They require people and companies to act, to abandon the old paradigms of the industrial and service economy to introduce new experiences and transformations, eventually resulting in a shift from an "experience economy" to a "transformation economy".
>
> *(Pine and Gilmore 2013: 40)*

This is a relevant experience in attracting Gen Z. According to a report (JWT Intelligence and Snap Inc 2019), this group, ranging in age from 13 to 22, has defined a new era of influence and creative communication.

According to the study, they are hyper-connected, have strong opinions, transfer their activism to the Internet and social networks and are also concerned about global problems and events. When asked about a slogan that would define generation Z, interviewees aged between 13 and 22 suggested variations of "be yourself" in first place, and in second place came variations of a focus on social responsibility like "save the planet", "we are the change", "we are the future" and "there is no planet B".

This suggests that in terms of sustainability, it is possible to explore socially themed activities and experiences which may require more active audience engagement, urging individuals to take action for a better world.

Another important aspect is how to use information. The Climate Outreach Charity, in partnership with the Global Call for Climate Action (a diverse network of more than 450 non-profit organizations in more than 70 countries), conducted research (Corner et al. 2015) to measure the best way to use images to communicate climate change. They conducted a series of informal interviews with academics and practitioners from the United Kingdom, Europe and the United States and focused on the most important questions for climate change communicators, according to existing literature and best practice.

The study resulted in seven key findings to raise awareness, such as the need to show real people, in unstaged scenarios, in communication with the public; that photography alone tells a story, contributing to the cognitive adaptation of the public with respect to the representation of climate change; that it is necessary to exemplify reactions to the scale of climate change on the daily routines of the public and that it is essential to present solutions to major climate tragedies, and not the act itself alone.

The study also points out the importance of showing local climate impacts to facilitate understanding of the concept; about being careful when using images of climate protests – that can give the idea that this is not an imminent problem, but something far away and, finally, the research suggests listening to the public response when applying different visual communication strategies about climate.

Museums in action in South America

According to the Register of Ibero-America Museums (Iberomuseos 2021), 16 Latin American nations are home to 8,795 museums. Focused solely on South America, the register shows that Argentina, Bolivia, Brazil, Chile, Colombia, Ecuador, Paraguay, Peru and Uruguay have a combined 6,785 museums, with Brazil home to the highest number of institutions (3,807). This reality justifies the higher number of Brazilian actions cited in this chapter compared to other South American nations. Specifically, innovative examples from Peru and Argentina will also be listed. All the projects presented herein are intrinsically related to the subject of climate change.

Museum of Tomorrow

Since its inauguration in 2015, in Rio de Janeiro, the Museum of Tomorrow has been offering digital experiences and has already welcomed over five million visitors. It was declared the most visited museum in Brazil by the Brazilian Museums Institute (Canonico 2017). The concept behind the museum is that its activities are grounded in the triad of science, art and technology: science creates audience engagement; technology permits the development of skills and art is able to evoke empathy, a characteristic developed when someone identifies what lies in the consciousness of another person, motivating them to do good (Bloom 2018). This movement provides an incentive for the creation of projects geared towards popularizing narratives that use art, technology and gamification as raw material. Gamification involves using online experience design and game mechanics to motivate and engage people so they attain their objectives (Burke 2015).

The Museum of Tomorrow offers a narrative about how we can live and shape a journey towards possible futures from the big questions that humanity has always asked, such as: Where do we come from? Who are we? What are we? Where are we going? How do we want to get there? According to the project of the Museum of Tomorrow (2015), developed by the Roberto Marinho Foundation, the Museum is considered a third-generation science institution that, with the assistance of audiovisual resources and interactive installations, encourages the public to look towards the next 50 years.

To understand this classification, it is important to look at the concept of science museum generations. McManus (1992) classifies the transformation that these institutions have undergone over the centuries. According to her, there are three types of science museums: the first generation, which consists of natural history museums; the second generation, composed of science and (technology) museums and the third generation, which highlights scientific phenomena and concepts.

Manso (2018) researched the new exhibit concept proposed by the Museum of Tomorrow and postulates that the third generation of science museums differs remarkably from the previous generations, as its fundamental principle is to explore scientific ideas and concepts. In the case of the Museum of Tomorrow, the narrative inspires a reflection on the symptoms of the new geological epoch, the so-called Anthropocene, in which humans become a geological force capable of altering the climate, degrading biomes and interfering with ecosystems. This is also the theme of the main permanent exhibit, which explains how humanity has brought the planet to this new level, demonstrating scientific evidence from research to date. The museum's content also invites visitors to reflect on whether they are willing to face the negative consequences of this new geological age, should we continue on this course or whether people are willing to work towards a better future.

At the Museum of Tomorrow, during the period from 2015 to 2021, exhibits were focused on telling stories about real-world events from a science and sustainability point of view, employing updatable data and facts. These powerful communication actions were one of the main ways to transmit knowledge to a diverse audience, having been developed in an interdisciplinary character, allowing for the presentation of subjects from different perspectives. These narratives showed that 84% of the public interviewed in surveys said they would be willing to review habits after having visited the exhibits – while among youths aged 16 to 24 years, the willingness to make change is 90% (Museum of Tomorrow 2020).

In 2019, the Food for Tomorrow – Feeding 10 Billion exhibition presented spoilers about our future: there will be many more of us, and in order to live well and coexist in urban spaces, we will have to change our habits, explore new agricultural frontiers and invest in technology. The exhibit showed visitors, for example, real food production projects in places affected by climate change, such as the Arctic tundras. In this place, where the average temperature is −10°C – even in warmer months – a rise of up to 5°C is forecast over the next 30 years, according to scientists (IPCC 2021). By showing visitors that climate change is at a point of no return – and that adapting to this phenomenon is essential to the survival of humanity – the exhibit was classified by audiences as informative and insightful. Over 251,000 people visited Food for Tomorrow over a period of six months, and 80% of the participants stated they were willing to review behaviour related to food (Museum of Tomorrow 2019).

The Fruturos – Amazonian Times exhibition, inaugurated in December 2021 and also shown over six months, presented to its 200,000 visitors all the grandeur, biodiversity and knowledge present in the Amazon. Produced over a period of four years, encompassing at least three research trips to different parts of the tropical forest including deforested areas overrun by agriculture, the exhibit used time as a narrative, presenting settings with interactive activities, elements that reveal the diversity and ambient sound of the region. Following the visit, at least 75% of the

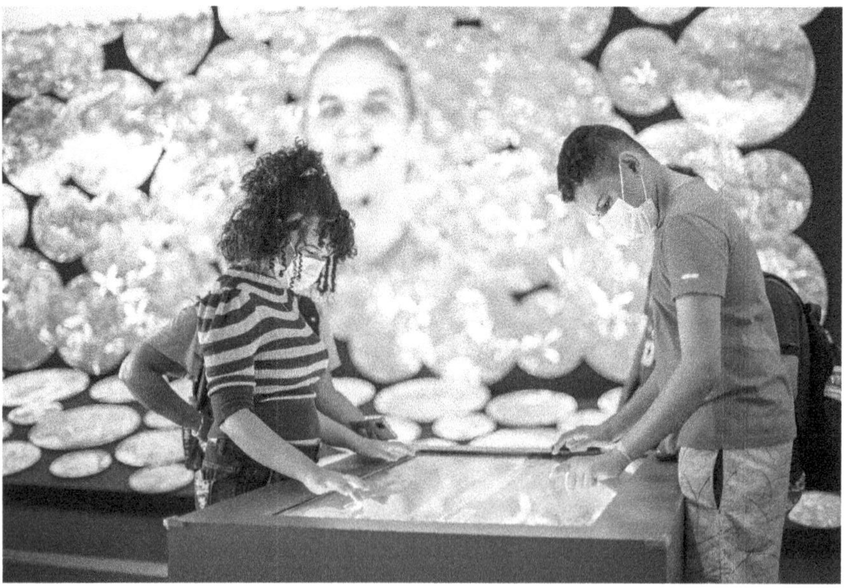

FIGURE 7.2 Photo from the exhibition Fruturos – Amazonian Times at the Museum of Tomorrow in 2022, Rio de Janeiro. This show reached 200,000 visitors. Photograph from the Museum of Tomorrow

public said they felt motivated to find more scientific information on the subject (Museum of Tomorrow 2022).

Also, innovative experiments provided data about the visitors' behaviour related to the themes presented in their exhibitions. The IRIS+ experience, which used IBM's artificial intelligence Watson, was developed to question visitors and inspire them to think about their role in society and truly participate in the search for more awareness, tolerance and a sustainable tomorrow (Candello et al. 2020).

A Watson Conversation Service (WCS) API (Application Programming Interface) hosted in the IBM Cloud was used to develop the application. The experience was trained to respond to doubts and also pose questions, guiding a dialogue with visitors with regard to the two main thematic branches of the Museum of Tomorrow: sustainability and coexistence.

The visitor's initial interaction with IRIS+ is through a voice dialogue. The conversation system guides this discussion. The dialogue begins with a question from IRIS+: "Considering what you saw in the main exhibition, what are you most concerned about?" Visitors then reply, and IRIS+ provides follow-up questions based on the response.

At the end of a conversation, IRIS+ recommends some social initiatives connected to the concerns mentioned by the visitor. IRIS+ has a database of previously registered efforts, and a recommendation component is responsible for identifying up to three social initiatives that are consistent with visitors' concerns. After being presented with recommendations, visitors can opt for a photograph. This photo is projected on a large wall, highlighting the most relevant themes for visitors, revealing clusters of people concerned about the same issues.

By February 2020, IRIS+ had been used 528,923 times. According to a survey by the Museum research team, of all the participants, 37% were engaged in an initiative related to their concerns about the planet. This data revealed that visitors seem more likely to act on issues as part of their daily routine, like environmental conservation and waste management.

The *New York Times* called IRIS+ one of the most disruptive uses of artificial intelligence in the world (Levere 2018). Its development has been the subject of congresses throughout Brazil, as well as events linked to technology held in France and the United States.

MuBE

The Brazilian Museum of Sculpture and Ecology (MuBE) is located in the city of São Paulo and was one of the first museums to declare a Climate Emergency. Conceived in 1986, following protests by residents calling for the preservation of green areas in an up-market region of South America's largest city, MuBE was designed by renowned architect Paulo Mendes da Rocha, who created the framework with the idea of water recycling already in mind and the avoidance of using major scenographies for exhibitions. In 2021, the museum opened an exhibition called "Por um sopro de fúria e esperança" (for a breath of fury and hope, in English), which approached the need for a rapid shift to a clean economy, without forgetting the need to bolster the resilience of the population that would be most affected by adverse weather conditions.

With the exhibit, inaugurated during the 2021 COP26 Climate Conference in Glasgow, UK, MuBe also launched its "Manifesto on the Declaration of Climate Emergency", with scientists linked to the IPCC, executives from major companies and artists included as signatories. Velloso (2022), the institute's director, affirmed that the action was a means to bring the public closer to the discussion and impacts how the museum is run. There is an effort to implement a circular economy in the production of exhibitions, with the reuse of settings, the acquisition of biodegradable materials and avoiding the use of MDF, which is substituted by the framework of the building becoming part of the exhibit design.

São Paulo Biennial

The last edition of the São Paulo Biennial, held in 2021, raised the issue of climate change through the voices of indigenous peoples. Under the theme "It's dark but I sing", a verse by Amazonian poet Thiago de Mello from the poem "Madrugada Camponesa", written by the author in 1965, the Pandemic Biennial brought together more than 1,100 works by 91 artists from all over the world, nine of them indigenous, presenting works on contemporary conflicts faced by different indigenous peoples, including deforestation. One of the highlights was artist Jaider Esbell, an indigenous member of the Macuxi people from the Amazon who died in 2021, whose work combined painting, writing, drawing, installation, and performance.

MAM Rio de Janeiro

In partnership with the Natural History Museum in London and the United Nations Museum, UN Live, the Museum of Modern Art in Rio de Janeiro is part of the program "My body: My planet", focused on engaging young people aged 12 to 18 years in a debate about climate change and global health. As Veras (2022), responsible for the project in Brazil, notes, the programme takes place in dialogue with museums in China, India and Nigeria, funded by the Wellcome Collection. In Rio de Janeiro, activists, students and university professors discussed possibilities for creating public action that unites the arts and sciences in favor of fighting deforestation, using Brazil's Carnival, one of the nation's leading popular culture manifestations, as an instrument to spread the word. The project resulted in a performance by artist Rafael BQueer, from Pará State in Brazil, in partnership with Argentinian designer Emilia Estrada, provoking the public to think about how they could collaborate to reduce the impact of climate change in the country.

Museo Moderno – Argentina

Since April 2022, the Museo Moderno, located in Buenos Aires, Argentina, has been presenting exhibitions as part of the "A Day on Earth" programme, conceived during the pandemic, when museums were forced to close their doors. It also considered a response by Argentine artists to the need for society to reflect on humanity's present and its relationship with the planet. Between 2022 and 2023, the institution presents 11 group exhibitions that relate the pain and suffering of climate-vulnerable populations to traumas suffered by the Argentine population, such as the years of military dictatorship – considered one of the bloodiest in Latin America, and which to this day mobilizes the population due to 30,000 missing people – as well as racial and gender-related violence (Museo Moderno 2022).

Voces por el clima Ecological Park – Peru

In 2014, COP20: United Nations Climate Change Conference was hosted in the Peruvian capital of Lima. UN signatory nations took the opportunity to continue the discussion on how to curb rising global temperatures, which have caused increased extreme weather conditions around the world. The space that hosted diplomats, activists and journalists was transformed a year later into Voces por el Clima (Voices for the Climate) Ecological Park, a 5,000 square metre area that houses exhibits about climate and its effects on Peruvian biomes and cities.

According to the Ministry of Environment from Peru (2022), five pavilions were erected using sustainable materials such as bamboo and wood, employing natural lighting and ventilation and featuring experiments and exhibits on forests, the ocean, energy, sustainable cities and mountains. Visitors are required to bring two plastic bottles as cost of entry into the park, which are recycled later.

The Inter-American Development Bank, responsible for promoting regional trade integration in Latin America and the Caribbean, classified the space as the first thematic park on Climate Change in Latin America (2016) and considers it an active agent to raise population awareness in facing the climate crisis with greater integration between public power and civil society.

Final remarks and further work

South America has an important responsibility to engage its population in tackling climate change because it is the "home" of the biggest tropical forest in the World. The Amazon Forest is spread across eight countries (Bolivia, Ecuador, Peru, Colombia, Venezuela, Guyana, Suriname, Brazil) and an overseas territory (French Guiana). Brazil is home to around 60% of the Amazonian territory, where all these ecosystems are present. These countries and ecosystems harbour an immense diversity of species.

Museums located in this part of the world are trying to do their best while there is a blackout in public policies regarding the protection of the forest and its people. It's important to mention that increasing far-right policies under Jair Bolsonaro's presidency in the most important country of the continent contributed to cuts in funding for cultural activities in Brazilian museums, reducing the sponsorships via a national law called the Rouanet Law. Some researchers called these actions a way of censorship (Molinero 2019). In October of 2022, the national election re-elected former president Luiz Inacio Lula da Silva to the presidency of Brazil, under the promises to recreate the Ministry of Culture and to act efficiently against Climate Change, protecting the Amazon Rainforest, for example. Therefore, there is now an ample opportunity to discuss how the culture could support engagement on this issue.

Knowing that climate change has a substantial impact on the health of biodiversity, cities and society, these small number of examples relating to arts, culture and climate education show us that there is hope, awareness and strength to call audiences to action using storytelling and supporting different ways to spread the message about our common house and about the future we want to build together – and most important, fighting against negationists and misinformation.

References

Abramovay, R. 2019. *Amazônia: por uma economia do conhecimento da natureza*. São Paulo: Abong; Edições Terceira Via; Iser Assessoria.

Bloom, P. 2018. *Contra la empatía*. Mexico City: Penguin Random House Grupo Editorial México.

Burke, B. 2015. *Gamificar – Como a gamificação motiva as pessoas a fazerem coisas extraordinárias*. São Paulo: DVS Editora, Kindle Edition.

Buster, B. 2013. *Do Story – How to Tell Your Story so the World Listens*. London: Do Books Company.

Candello, H., Barth, F., Carvalho, E. and Cotia, R.A.G. 2020. Understanding How Visitors Interact With Voice-Based Conversational Systems. Design, User Experience, and Usability. *Design for Contemporary Interactive Environments*, 40–55.

Canonico, M. 2017. *Em seu 1° ano, Museu do Amanhã se torna o mais visitado do país*. https://www1.folha.uol.com.br/ilustrada/2017/01/1854076-em-seu-1-ano-museu-do-amanha-se-torna-o-mais-visitado-do-pais.shtml (Accessed 6 October 2022).

Corner, A., Webster, R. and Teriete, C. 2015. *Climate Visuals: Seven Principles for Visual Climate Change Communication (Based on International Social Research)*. Oxford: Climate Outreach. https://climateoutreach.org/reports/climate-visuals-seven-principles-for-visual-climate-change-communication/ (Accessed 6 October 2022).

Iberomuseos. 2021. *Registro de Museus Ibero-Americanos*. http://www.rmiberoamericanos.org/Home/Recursos (Accessed 1 October 2022).

Inter-American Development Bank. 2016. Voces por el Clima: El primer parque temático sobre cambio climático en América Latina. *Ciudades Sostenibles*. https://blogs.iadb.org/ciudades-sostenibles/es/voces-por-el-clima/ (Accessed 1 October 2022).

IPCC. 2021. *Climate Change 2021: The Physical Science Basis*. IPCC Sixth Assessment Report Working Group I: The Physical Science Basis. https://www.ipcc.ch/report/ar6/wg1/ (Accessed 1 October 2022).

Ipsos. 2022. *Climate Change: Severity of Effects and Expectations of Displacements.* https://www.ipsos.com/en/climate-change-effects-displacements-global-survey-2022 (Accessed 13 October 2022).

JWT Intelligence & Snap Inc. 2019. *Into Z Future.* https://www.wundermanthompson.com/insight/into-z-future-understanding-generation- (Accessed 5 October 2022).

Levere, J.L. 2018. *Artificial Intelligence, Like a Robot, Enhances Museum Experiences.* https://www.nytimes.com/2018/10/25/arts/artificial-intelligence-museums.html (Accessed 5 October 2022).

Lewrick, M., Link, P. and Leifer, L. 2018. *The Design Thinking Playbook: Mindful Digital Transformation of Teams, Products, Services, Businesses and Ecosystems.* Hoboken, NJ: Wiley.

Manso, B.L.C. 2018. *Museu do Amanhã: Uma nova proposta de museu de ciência?* (PhD thesis Federal University of Rio de Janeiro: Instituto Brasileiro de Informação em Ciência e Tecnologia).

McManus, P.M. 1992. Topics in Museums and Science Education. *Studies in Science Education*, 20(1): 157–182.

Ministry of Environment from Peru. 2022. *Voces por el Clima.* https://www.minam.gob.pe/vocesporelclima/pabellones-tematicos/ (Accessed 5 October 2022).

Molinero, B. 2019. *Cultura, sob Bolsonaro, vive volta da censura, perda de ministério e viés evangélico.* https://www1.folha.uol.com.br/ilustrada/2019/12/cultura-sob-bolsonaro-vive-volta-da-censura-perda-de-ministerio-e-vies-evangelico.shtml (Accessed 10 October 2022).

Museo Moderno. 2022. *Un día em la tierra.* https://museomoderno.org/exposiciones/un-dia-en-la-tierra/ (Accessed 5 October 2022).

Museum of Tomorrow. 2015. *Plano Museológico do Museu do Amanhã.* https://museudoamanha.org.br/sites/default/files/expomus_planomuseologico_digital_160219_Otimizar.pdf (Accessed 29 September 2022).

Museum of Tomorrow. 2019. *Pratodomundo – Comida para 10 Bilhões.* Rio de Janeiro: Museu do Amanhã.

Museum of Tomorrow. 2020. *O Amanhã do Rio.* Rio de Janeiro: Museu do Amanhã.

Museum of Tomorrow. 2022. *Pesquisa Fruturos – Tempos Amazônicos.* Rio de Janeiro: Museu do Amanhã.

Pine II, B.J. and Gilmore, J.H. 1998. Welcome to the Experience Economy. *Harvard Business Review*, 76(4): 97–105.

Pine, B.J. and Gilmore, J.H. 2013. The Experience Economy: Past, Present and Future. In Sundbo, J. and Sorensen, F. (eds) *Handbook on the Experience Economy.* Cheltenham: Edward Elgar Publishing Ltd, 21–44.

Velloso, F. 2022. *Mube and Climate Emergency.* Interview by Eduardo Carvalho [Google Meets], 16 September.

Veras, L. 2022. *My Body: My Planet in Brazil.* Interview by Eduardo Carvalho [Google Meets], 6 September.

8

COLLECTIONS MANAGEMENT AND CONSERVATION

Caitlin Southwick

Introduction

The climate crisis is caused by humans and exacerbated by certain behaviours. Some of these behaviours are obvious – such as consuming energy derived from fossil fuels – and others are more subtle – such as using single-use plastics. And more complex are behaviours that are believed to be sustainable but are not (as a consequence of buying into greenwashing).

The cultural sector has unique behaviours, issues, values, and circumstances related to sustainability. As a sector built on perpetuating knowledge, culture, beauty, history, and societal cohesion, museums are naturally inclined towards social sustainability. However, until recently, issues of environmentalism were less associated with a sector focused on art and artefacts.

While the core values of the cultural sector remain – to protect and house cultural heritage and to serve our audiences and communities – the recently clarified connection between social justice and climate justice brings the climate crisis to the forefront of museum practice. Museums must now question their role in the larger societal structure: how they are contributing, mitigating, advocating, and adapting to climate change.

In light of current societal crises, the purpose of museums and the role of museum professionals is changing. Yet, while many individuals see the need for change and want to take action, the cultural sector has been slower to engage. Systemic inertia, financial insecurities, and hyper-productivity are some of the reasons that change has been slow to start. However, that is rapidly changing.

Systemic barriers

Barriers for sustainable engagement are the same at many institutions – no matter the size, what type of organisation, or where in the world they are located. While

DOI: 10.4324/9781003347606-9

many institutions cite issues related to their particular circumstances, these issues are actually systemic. The sector, as a whole, is financially unsustainable, suffers from short-term thinking, lack of transparency, and an obsession with professionalism. There is an unsustainable dependence on productivity and ideas of professionalism which produce unrealistic expectations and goals and, at the same time, an inertia that seems to exemplify the Dutch phrase "als je boven het maaiveld uitsteekt, dan val je op." Loosely translated, this infers that it is not a good thing to stick out. No one wants to be the first to do something.

These barriers have made it difficult for the sector to make change quickly (Weisberg 2023). However, being aware of what the challenges are may enable museums to think about our priorities and start thinking about what our role is in the climate crisis and how we collect and display cultural heritage and art.

Greenwashing, art washing, and green hushing

In many sectors, greenwashing is one of the most dangerous forms of unsustainable behaviour. Examples include companies putting "green" labels on their products, which are unsubstantiated or fossil fuel companies promoting their renewable energy investment to detract from continued drilling and increasing emissions (Vetter 2021). Greenwashing, defined as "behaviour or activities that make people believe that a company is doing more to protect the environment than it really is" (Cambridge English Dictionary 2022), is less prevalent in the museum field. However, green hushing, possibly an even more dangerous pitfall, is common. Green hushing "refers to a company's refusal to publicise anything about the sustainability of its products or services for fear of being open to criticism" (Stolle 2022). Museums can easily and, unknowingly, be green hushing. Museums' fear of not being perfect or of upholding the highest "standards" of professionalism in combination with behind the scenes questionable practices or financial links means that they often don't publicise what they are doing well for the fear of being called out for what they are doing wrong. For example a museum which is working on reducing its carbon footprint by upgrading its building systems and equipment, increasing ranges on climate control, and installing renewable energy systems may be reluctant to discuss these impressive measures if they accept money from a fossil fuel company (the author will not cite specific references for this in order to not shame or blame any specific institution).

And while green hushing is something that may not get as much attention by the public, there has been a lot of discussion about divestment from fossil fuel companies and the issue of "artwashing." According to Creative Carbon Scotland,

> "Artwashing" is a term that has been popularised to describe the ways that major polluters and other organisations that cause environmental damage might use the arts to clean up their public image. The term has been at the centre of recent debates about issues like sponsorship from oil and gas companies in the

> arts, fossil fuel publicity methods and how the arts might unknowingly be used as a way to obstruct progress on climate change.
>
> *(2022)*

Many feel strongly that museums and galleries should avoid accepting money from companies, individuals, and organisations who don't share the same values as them, which calls into question why a museum would consider aligning themselves with these companies in the first place, and what the ramifications are for the influence of these external funding sources on how the museum functions. And while the sentiment is there, the transparency is not. This will be examined further later in the chapter.

Museums may be surviving in constant fear of budget cuts or losing support and funding (Lindqvist 2012). This fear puts many organisations into survival mode and prohibits change and even questions mission fulfilment.

Funding

The reason behind accepting large sums of money from disreputable sources is simple: financial unsustainability. Culture often sees the first budget cuts, and there is a constant battle to attract funding or prove that the investment is of value (Altamimi and Liu 2022). The fear of course is that beggars can't be choosers, and if the museum loses funding because of their principles, where will the replacement funds come from?

Museums, by ICOM's definition, are not-for-profit organisations (ICOM 2022). In many countries, this inhibits sustainable financial models and forces a reliance on grants, donors, and government funding. Museums which rely on public funding or philanthropy (as opposed to self-funded) are subject to the influence and control of these funding bodies. If a donor or a government is not on board with climate discussions, it can severely affect what a museum is able to produce, display, or the programming it can create. Museums who have made the choice to accept money from sources that don't align with the museum's published values may experience inhibited progress with social and environmental issues, and museums may have additional reservations or restrictions about their activities and responsibilities because of the source of their funding.

Some museums cite political reasons for not engaging with dialogue on climate (Gill 2022). State-run museums in countries such as Canada and Hungary, where the local or federal government is not aligned with the climate crisis, feel they cannot talk about issues that don't align with the current government agenda. This extends to private funding. Museums in the United States may have board members or trustees who feel climate is not a conversation for a museum, therefore inhibiting the museum from working with this issue. While this is a topic that is widely discussed and commonly known in the field, it is not publicised and conversations tend to stay behind closed doors, further exacerbating the problem.

In some situations, museums can be more enterprising, finding ways to be financially independent. But sometimes these measures can receive negative backlash or call into question the museums' values or practices, as seen during the pandemic when museum directors were considering selling off collections to pay salaries for employees (Pogrebin and Small 2022). It is not needed to mention the stark value conflict here: the mission of the museum to collect objects versus the well-being of the staff. In a sustainable sector, there would be no need to make such a choice. The few museums, however, that have found ways to bring in self-sustaining and ethical funding are generally the institutions that have had the most success in achieving sustainability progress. A prime example of this is the editor's institution, the Horniman Museum and Gardens. However, even these institutions, in the author's experience, are still citing budget restrictions as bottlenecks to sustainable action. When presented with solutions for sustainability, the first response is that there simply isn't funding or budget for investing in sustainable activities.

Even in circumstances where values align, or where museums are making their own money, budget is still cited as the biggest barrier to change. The problem is that budgets will not change unless the sector rethinks its business model – and its commitment to sustainability.

Short-term thinking

The fear from financial unsustainability often leads museums to look to short-term thinking and think that they cannot afford economic or social sustainability. These are two fundamental issues when it comes to change in the sector. Because as long as institutions have this short-term mindset and don't embrace sustainability, they will continue to be kept in the cycle of budget restraints, dependence, and influence from external sources and lose their position of trust and relevance with the public.

Short-term thinking is dangerous for a sector which is predicated on the long-term preservation of human heritage. The obstacles this habit imposes are limited to not only the outputs of the museum (such as the influence on content discussed earlier) but also operations. If a museum is financially unsustainable, it makes it difficult to be environmentally or socially sustainable internally. A museum which is focused on surviving may not see the benefit of investing in longer term projects, even if there is a significant financial benefit. Sustainability has the assumption of being expensive. Upgrading an HVAC system, switching to LED lights, or installing a heat pump, for example, may require upfront investment. Even though these measures will pay off in the long run (which they all do), the short-sighted financial lens makes them seemingly unachievable for many. The mentality that we don't have the money now, so we won't invest in our future seems slightly contradictory for a sector whose purpose is to house and care for humanity's history and culture in perpetuity.

Additionally, these short-term pressures are reflected in staff and work output. Temporary and touring exhibitions, which are in many cases relied upon for revenue generation, result in short-term contracts and an emphasis on mass production. An institution may try to put out 12 exhibitions in a year so they can stay relevant in the press and then ship an exhibition all over the world to generate income. However, this model puts tremendous pressure on the staff, not to mention the carbon footprint and waste produced (see Artists Commit: https://www.artistscommit.com/ for carbon footprint assessments of exhibitions).

Hyper-productivity

To continue to obtain funding and to stay relevant, museums may try to do more: put on more exhibitions, acquire new things, and bring in bigger objects from faraway places. Basing metrics of success on numbers – larger audiences, more ticket sales, more things, more exhibitions – is aligning museums with consumer capital ideals (Anderson 2004: 5). But by redefining success, the cultural sector can do better. Using the principles of degrowth (although perhaps using less alarming semantics), value systems can embrace the "voluntary societal shrinking of production and consumption aimed at social and ecological sustainability" (DeMaria et al. 2013: 192). Instead of valuing how many people come through our doors or how many exhibitions we can put on in one year, we can focus on what the experience is, what the impact is, and what people leave with (and not just a plastic Van Gogh tote bag).

To rethink hyper-productivity and metrics based on the capital values of the business world, we have to slow down and re-evaluate our missions and our values. Slowing down can start with production. Instead of doing seven exhibitions a year, can we do three? Can we measure not how many people come to an exhibition, but how long they stay and how they connect with the topic? Can we stop measuring in financial terms and instead focus on social impact? This may be more difficult to sell to some investors, but then are they really the investors we want. This goes back to our financial sustainability and the previous discussion about the influence of funders on our work.

To be sustainable, we must examine the fundamental questions: Why are we doing this? Who are we doing this for? What is the value, the purpose, the mission of a museum? Is it just a production house for exhibitions? Is that what we have become? If museums were created to be houses of collections, their role in society has changed. The new museology, which puts people at the heart of culture – from our practices to how we define what is successful – demonstrates a model for museums which embraces slowing down, valuing people over production, and social sustainability (Desvallées and Mairesse 2010).

New museology

To understand the future of museums, it is important to remember how they have developed over time and around the globe, how they have shaped society, and how society has shaped them. History teaches us lessons to learn and not to repeat – if we are astute enough to listen.

Historically, collections care has been at the heart of the museum. As stewards of humanity's heritage, museums were tasked with caring for and displaying the objects that comprise the history of humanity and nature. Museums began as spaces for collecting rare objects and artefacts, a place for intellectuals to gather and exchange knowledge and experience, a place for education. Stemming from the original cabinets of curiosities, museums were fundamentally places for gathering knowledge and preserving objects (Richman-Abdou 2018). However, over the past decades, that role has changed. And in light of the climate crisis and social justice, it is changing again.

New museum theory places the audience at the centre of a museum's purpose, rather than the collection: "new museums are ecomuseums, social museums, scientific and cultural centres, and generally speaking, most of the new proposals aimed at using the local heritage to promote local development" (Desvallées and Mairesse 2010: 55). According to new museum theory, museums are now not only repositories of knowledge and objects but socially engaging, community-facing entities as well. This new social aspect changes not only what museums do but also what they communicate, how they communicate, how they practise, and who they are for. Collections care is no longer the core activity of a museum, but a complementary asset to the messaging and representation of the past and, more interestingly, the future.

As this shift progresses, the concept of collections care must adapt to incorporate other, more pressing topics, including social justice and climate change. And this also means that museums need to look at their priorities and the way they are managing their collections and their current relevance, which, in this author's opinion, is linked to their sustainability.

If the "new museology" has been around since the 1970s, why is this not mainstream practice? (*ibid.*). Why do we continue to rely on this unsustainable system which keeps us in this vicious cycle of lack and limitation, rather than valuing our staff, our audiences, and our communities? Why is sustainability not the number one priority? And while the new museology promotes a lot of very interesting and sustainable values, it doesn't include the climate crisis. Environment is not explicitly mentioned in the ICOM Museum Definition, instead incorporated into the more general term "sustainability," which is open to interpretation (ICOM 2022). Environmental responsibility as a foundational element of a museum's purpose is still largely missing. And we are not talking about it.

Transparency

Lack of transparency hinders our abilities to become sustainable. It is a common adage that you can't manage what you can't measure, and this is true not only for calculating and reducing carbon footprints but also for all aspects of sustainability – and for maintaining our positions in our communities. If this lack of transparency in our actions and our partnerships continues, museums are at risk of losing the public's trust. Conversely, if we are open and transparent about what we are doing well and what we need improvement on, we may find that the answers can come from unexpected sources. This is the idea behind co-creation and belonging. If a museum is trying to reduce its waste, it may rethink what happens to the materials after the exhibition. If the museum is working in a silo to try and solve this issue, it may never find a solution. But if the museum advertises that it has materials in excess and invites the community to help solve the problem, partnerships and opportunities may arise. The Denver Art Museum works with a local organisation, the Resource Area for Teaching, to find new life for materials that otherwise would have been taken to landfill (New York Times 2014). Instead of hiding unsustainable practices, this type of creative thinking demonstrates the community spirit of the museum. Inspiring our audiences and our communities with our success and asking for help and support in solving problems we are currently facing contribute to a sense of belonging: our communities feel they can support us as we support them. This is sustainability at its finest.

This idea of transparency should apply to all aspects of a museum's organisation, including areas where museums feel they are not doing as well as they should. There are many ways in which museums can be more transparent – from financial streams to health and safety issues. There is transparency as an institution – with audiences and peers – and transparency within the institution – with staff and internal communication. Many systemic issues regarding sustainability (carbon footprint, diversity and inclusion, accessibility) stem from a lack of transparency – inside or out.

For example in the second half of this chapter, the author will explore issues of conservation and collections management, with an emphasis on climate control. This specific example demonstrates all the barriers mentioned earlier: particularly short-term thinking, professionalism, and transparency. As will be discussed, museums are often requiring strict conditions for loans that they don't maintain themselves. If museums were to be transparent and open about what conditions they actually keep their collections in (not to mention considering what conditions are actually needed), then the problem of loan agreements being prohibitive clears up immensely. This fear of transparency is particularly tied to the addiction to professionalism (or rather the appearance of it) in the field. The author suggests, in this particular instance, we change the jargon from "relaxing" standards to "updating." That (hopefully) would solve at least the appearance of professionalism for climate control. But unfortunately the "professionalism standard" is not related to climate control alone. Once certain practices have been adopted, it is often difficult to revert to older practices or to change to other not as widely implemented practices. Again, this goes back to the expression about sticking your head out of the cornfield.

Transparency (or rather a lack thereof) is fear-based – not only a fear of looking unprofessional, or of negative repercussions (as was mentioned earlier related to green hushing), a fear of losing funding or of getting shamed for where funding is sourced, but also a fear related to job security. This transparency is not just lacking with external and public-facing communications, but also with internal communication and staff relations.

Staff are often siloed, not only within institutions but also within departments. Hindrance to change can sometimes be as simple as not having a conversation with your colleague. One colleague may assume that another would be averse to change and therefore never asks the question. Or a staff member may feel uncomfortable approaching upper management and questioning the status quo. Institutions where there is transparency in the hierarchy (great examples are the Anchorage Museum in Alaska and the Baltic Gallery in the UK) allow staff to drive change because they are able to have transparent and open conversations with decision-makers.

If this transparency is not established within an organisation, change is often stalled. Sustainability must be simultaneously top-down and bottom up. If a Director is on board, then they may not have the capacity to change the whole institution on their own (after all, they still have to run an institution), but the bottom-up approach is almost never successful if there is no buy-in from leadership.

Professionalism

The museum world has always been rooted in academia, and the sector has a certain prestige. The museum professional often is well-educated, from a certain upbringing (after all, individuals who get into the museum profession are more likely those who have had the privilege of being exposed to museums), and professionals who become museum practitioners are expected to maintain a certain level of professionalism. Tate did an interesting expose into the aspect of professionalism in the field in 2005 which explores what professionalism means and how it shapes the work of the museum practitioner (Charman 2005). This analysis, which in the author's opinion remains largely true, covers the psychology or history of how museum professionals were shaped. What is interesting is how this relates to climate action in the field.

As is becoming a theme in this chapter, professionalism is linked to inertia. Professionals maintain the status quo and don't question or cause disturbance. And the climate conversation requires questions, it requires challenging current practices, and it may be seen as a disturbance. Once a certain practice or standard is implemented, that is the course followed by professionals, and changing that practice could call into question the standard of professional practice.

The climate crisis in the art world, particularly as of late, has been linked to activism. Groups such as Just Stop Oil and Stop Fracking Around have targeted museums for demonstrations – throwing soup or syrup on paintings (Gayle 2022;

Ditmars 2022). Museum directors responded with a declaration calling this act vandalism citing that the activists "severely underestimate the fragility" of the works and cite the "core tasks of the museum as an institution – collecting, researching, sharing and preserving" (ICOM Germany 2022). This "professional" reaction polarises the climate conversation from museums even further.

Once people feel they are doing something in a "professional" way, it is very difficult for them to change their practice and justify this from a professional standpoint. Consequently, making changes for the sake of the environment might not align with the current professional standards of being a museum practitioner. However, in the author's opinion, it has to. Being a professional means caring not only about your job today but also about your work next month and next year and the legacy being left behind. And isn't that what museums are all about – ensuring that the legacy of humanity through culture be preserved for future generations? Consequently, the new professional standard must include adaptation, change, or upgrading to include environmental responsibility. Rather than hiding behind tradition, professionals now must be future-oriented.

Museum professionals, conservators, and cultural professionals in general tend to be conservative, promoting and maintaining very high standards of "professionalism" and academic integrity. This desire to uphold professional appearances tends to lead professionals to rely on the status quo, which inhibits change. Museum and gallery workers practise what they were taught, and many seek out advancement in their careers through academic publications, which rely on an antiquated and long drawn out peer-review process that results in research already being years old by the time it is published.

It is not only in the cultural sector that this idea of professionalism is hindering sustainable action. Aho (2012) describes a similar situation in the construction industry as well, "The property and construction industry's lack of progress towards sustainability is sometimes blamed on corporate short-sightedness and short-term profit-seeking." Aho goes on to explain that the business models value production as performance evaluation and that there are no reward systems in place for sustainable practice:

> [S]ustainable professionalism means consistency and integrity in applying one's skills and competencies for the benefit of the community. Unfortunately, in the current state of affairs this indeed is a matter of individual ethics, and progress is understandably constrained to a certain extent by the need for securing individual short-term economic survival. Only few inbuilt mechanisms exist in the industry for promoting, rewarding or incentivizing this type of individual behaviour.

This idea of success metrics based on performance, production, and professionalism (in the case of museums, often tied to academic integrity and prestige) often prohibits individuals from engaging in behaviour change.

Professionalism not only is a systemic issue, but it also applies to the individual.

Personnel

While museums are maintaining unrealistic standards and "professional" (outdated) practices, and simultaneously trying to adjust to the new requirements of a community-centred organisation, museums as institutions have been slow at taking up the discussion of the climate crisis (not to say that there aren't leaders and action in this realm – the California Academy of Science, the Manchester Museum, the Australian Museum, and others have integrated climate in a genuine and holistic way). Interestingly, this institutional inertia is met by a strong demand from the individual practitioner. Environmental action seems largely to be coming from a more grassroots approach. However, the same barriers apply at the micro level as the macro. Cultural professionals generally want to be sustainable, but systems are prohibitive.

These barriers take many forms from financial to work–life balance to health and safety – all related to fundamental principles of sustainability and systemic societal issues. While systemic issues require collective action and consensus, individual action often relies on another, maybe more complex requirement: courage (and perhaps better pay).

Museum professionals grasp the importance of environmental sustainability. Working with materials – whether that be in the context of designing and building an exhibition, the physical collections, or conserving them, this is often an entry point for people to see waste and unsustainable practice, and they want to do something about it. However, while there is this palpable desire to make a difference and take action, there is also a sense of helplessness and inability to do so. The plight of the museum professionals is that they are not, generally speaking, empowered to be sustainability leaders.

Individuals experience the same barriers that apply at the sectoral level. Short-term thinking on a systemic level results in short-term contracts for practitioners. Many professionals rely on grants or other temporary contracts. More and more work is being outsourced. Fewer museums have in-house conservators, and there are more freelance curators than ever (Indrisek 2018). This puts additional pressure on the amount of work that can be produced in such a short time, which results in hyper-productivity. If a museum professional wants their contract renewed, they have to do the best, and the author has observed this competition result in highly uncollaborative environments. Again, short-term thinking means massive pressure for the individual, who is subsequently also in survival mode – fighting for their job, where they are more likely already underpaid, and adding more to their plate to stand out or secure their position. Museum professionals often take on additional activities, such as volunteering for boards or committees for national or international organisations, or presenting at conferences – including on sustainability.

People work in the cultural sector because they are passionate about it, and they believe in the higher mission of culture. However, this love can lead to unsustainable practices. Jobs in the cultural sector are few and far between and often mean high competition, low pay, and unstable positions. "It would be difficult, if not impossible, to find a museum professional who got into the arts for the money. But even curatorial assistants have to pay rent" (Halperin 2017).

Museum practitioners are generally made to feel that they should be grateful that they even have their jobs. In an interview with a conservator (both the individual and institution will remain anonymous for their privacy), the author was told that the institution this conservator worked in refused to give her a raise and that she was expected to take on the role of the sustainability lead for free – and continue to fulfil all of the duties of her conservation position.

> They made me feel like I should be so lucky to be there, almost as if I should be thanking them for the chance to work at such a prestigious institution and therefore should take on the extra work at the same low pay with a smile on my face.

This particular conservator felt so undervalued, overexploited, and frustrated that she finally quit. She is now a full-time sustainability lead – and not in the cultural sector. It had taken her two years to establish a green team at her organisation, and all of that momentum left when she did. None of her colleagues had the capacity to take up the extra work. This is not to say they didn't care, but they just had too much on their plates already.

Museum professionals are generally underpaid, and even if the salaries were higher and fairer in the cultural sector, sustainability is still seen as an add-on – something outside of the daily roles and responsibilities of the practitioner. Again, this is not true for all. There are organisations the author is aware of that are looking at pay schemes for staff joining the Ki Futures programme and allocating a percentage of salaries to sustainability work. Additionally, the position of a sustainability lead or climate curator has been established at museums like the Bishop Museum in Hawaii, the Royal Ontario Museum and the Australian Museum. However, for the vast majority of organisations and individuals, activities related to sustainability are still seen as an add-on. And museum professionals simply don't have the time.

Hyper-productivity of the institution trickles down to hyper-productivity of the individual. The mentality that "more is better" results in unsustainable practices, such as being asked to produce multiple exhibitions every year, as previously explored. This resource-heavy practice also affects our mental well-being. To put years of time, energy, and creative thinking into an exhibition, only to have it shown for six months and then literally tossed in the trash is not only environmentally damaging, it is also emotionally shocking. Museums are about preserving our intellectual property, our tangible and intangible heritage, so why is the work of a curator less important than the artist who is displayed in the exhibition?

Financial insecurities and time constraints aside, many museum professionals might not feel that they have the agency to drive change at their institutions or don't want to risk losing their jobs by being controversial. Museum professionals not only are often underpaid and overworked but also have to maintain the same standards of professionalism that are associated with their institutions. This fear of seeming unprofessional (and feeling completely replaceable) propagates the inability or indecision to act. To question the status quo, to confront the work that we are doing from a fundamental standpoint, and think about reassessing our values and ensuring that our day jobs align with our moral and ethical compasses are climate work. And in a fear-based system, it is scary.

This entire system makes it incredibly difficult for cultural professionals to act. Professionals who are overworked, underpaid, and mentally spent (not to mention all of the other things on their plates – like families or personal lives) are unable to see how they can possibly take on one more thing. It seems that we don't have room for the ecological crisis because we are running in short-term, survival mode. And sometimes that is more literal than figurative. In the conservation field, the use of toxic chemicals has resulted in the development of cancer and even death in practitioners globally. Conservators who develop allergies to their work because of the use of hazardous materials such as toluene and bi-component epoxy resins can no longer continue their jobs. Even more dire are conservators who die of cancer caused by the exposure to toxic chemicals (the author is omitting specific references to protect the privacy of conservators interviewed). Instead of finding a healthier alternative in conservation work, they are forced into other careers – becoming collections managers or registrars, having to give up on their practice. Again, which is unsustainable.

And with all of these difficulties and fears that cultural professionals face every day – job loss, health and safety concerns, and the fear that someone else will come along who is more professional – we are actually hurting not only ourselves but also our field and, most importantly, the planet. We are paralysed by fear of losing our jobs.

> [Y]ears of pay freezes and budget cuts have had a severe impact on the living standards of museum workers. The pandemic and now the cost-of-living crisis have taken a toll on wellbeing and pay; as a historically underpaid sector we are falling further behind as inflation hits double digits and the cost of essentials soars.
>
> *The Museums Association 2022)*

This all, of course, means that while the ecological crisis should be at the forefront of our thoughts and our work, it often gets pushed to the back burner. We are in this constant state of fear and overwhelm, and that makes it difficult for us to take on more. And sustainability is more. Adding one more thing to our plate – without taking something off – is near impossible. We are already perilously placed with our mental health and our time, it is not reasonable to ask people to now take up the biggest threat faced by mankind. But at the same time, it is impossible not to. We need culture to tackle this, and we need every single cultural professional not only on board and caring but them actually taking action and making this their number one priority.

But again, whether institutional or professional barriers, these issues are not being widely and openly talked about. Perhaps for fear of looking unprofessional or for fear of deterring future funding, the issues of unsustainability in the cultural sector are widely underpublicised, underdiscussed, and generally accepted as a norm. The inertia of the institution may be saved by the action of the individual, if the individual is brave enough to take that risk.

Conservation and collections care

Overview

Perhaps the most pertinent issue today related to museums is the history, display/use, and care of their collections and how these aspects relate to audience engagement and interaction. It is still a core value of museums to care for and display collections. But collecting and collections care have become unsustainable practices: over-accessioning, showcasing contested objects, overuse of energy to maintain unrealistic and unnecessary climate control conditions, and colonised narratives propagating structural racism. These issues are at the heart of how museums must change to truly embrace sustainability.

While the author wishes to acknowledge the social implications around collections care – from decolonization and repatriation to accessibility and inclusion – the focus for this section will be on the environmental impact of collections care, particularly outlining issues around conservation and climate control. While this is a well-known issue, it is difficult to discuss, because in many circumstances, we don't actually know what the impact is.

Climate control

Climate control requires massive energy consumption, which equates to carbon emissions. However, to date, there is very little information available regarding the carbon footprint of museums and galleries. It is not yet common practice to measure (much less publish) carbon footprints of museum buildings. In the 1990s, governments around the world were alerted to the massive carbon impact of buildings, and benchmarking regulations were mandated in many countries around the world (e.g. the Energy Star Portfolio Manager, which was created in 1992 (EnergyStar 2023)). However, in some cases, through loopholes or shortages, cultural institutions were exempt from these requirements (Fan 2014). And there had been very little incentive or reason for museums to be taking it upon themselves to measure their own carbon footprint, putting benchmarking and general metrics 20 years behind other sectors.

In addition to little information about the carbon footprint of museums as a sector, as previously mentioned, museums don't typically make their climate control regulations public. This lack of transparency leads to unrealistic expectations from lending institutions and a paralysis for change, integrated in trying to uphold "professional standards" and missed opportunities for communication among global colleagues. If this is widely known as one of the most energy-intensive and inefficient practices in the field, then what is the reason we have not already addressed and changed this practice (Marshall 2023)?

Again, the same principles that apply to sectoral inertia for sustainability as a whole apply specifically to museum environmental control. These range from financial (fear of losing out on object loans, which is a main source of income), intransparency (museums not communicating with each other about practices), to professionalism (changing current conditions to less-stringent guidelines feels like lowering your standards), and time and financial pressure (not enough money to upgrade systems or pay for a preventive conservation specialist to support the transition and not enough time to do it properly). And of course this is all foundationally supported by short-sightedness. Upgrading climate control ranges to reflect collections needs and individual circumstances will save money in the long run. This is long-term cumulative gain. It may take some time or some upfront investment, but small changes can lead to big savings, as demonstrated by the Guggenheim Bilbao who saved €20,000 a *month on their energy* bill by upgrading their ranges (Marshall 2023). But short-sightedness in the sector makes taking the first step feel impossible.

The history

There are many publications on the history of climate control, and in December 2022, Ki Culture, in partnership with the Gallery Climate Coalition, presented a two-day event to outline the history and current scientific research about the issue (Ki Culture 2022). Therefore, only a brief summary will be presented here, emphasising where the miscommunications took place which have led to current practices.

In the 1960s, the concept of climate control for preventive conservation was introduced, with scientific data indicating the possibility of maintaining specific climatic conditions for material preservation (Plenderleith and Philippot 1960). At the National Gallery in London, 60% relative humidity and 60 degrees F was specified as ideal conditions for their paintings collection, according to Dr Harold Plenderleith. These parameters were not intended to be the best conditions for the National Gallery in Australia, or a natural history museum in Brazil. However, assigning controllable numbers to a previously uncontrollable parameter was enticing. This human desire to control, with the added prestige of the institution resulting in a label of "best" practice, had people rushing towards implementing these conditions. Museums worldwide installing new HVAC systems scrambled to adapt these very stringent ranges in their buildings and storage facilities, resulting in huge increases in energy consumption, without taking into consideration the origins or purpose of the published numbers or of their own needs, contexts, and collections.

However, over the last 50 years, new scientific data has emerged, proving that larger ranges are acceptable for collections care. Research, declarations, and guidelines have been published from the Bizot Group, the Australian Institute for the Conservation of Cultural Materials (Boersma et al. 2018: 12; AICCM 2018), the American Institute of Conservation, the International Council of Museums Committee on Conservation and International Institute of Conservation (ICOM-CC and IIC 2014), and ASHRAE (formerly named American Society of Heating, Refrigerating and Air-Conditioning Engineers). These guidelines, moving away from strict standards of +/− 2 or +/− 5, embraced considerations of historical conditions, seasonal or geographic changes, building capacities and installations, and wider ranges (+/−10).

Scientific research has proved that the strict climate "standards" implemented at institutions are unnecessary. Stefan Michalski and Gaël de Guichen, two pioneers in preventive conservation including environmental management, have published widely on the topic. At the 2018 IIC conference on preventive conservation, de Guichen points out how various materials have lasted hundreds of years in self-regulating conditions and that subjecting objects to human controlled climates could actually inflict damage on objects (de Guichen 2018). Current research concurs that climate control at stringent levels is unnecessary (Boersma et al. 2018; Michalski 2007; Ki Culture 2022). However, the transition to these new, less stringent regulations has been slow on the uptake. Johnathan Ashley Smith states, "It is surprising in an age of instant global communication how slow and patchy the spread of new knowledge can be" (Ashley-Smith 2018).

While this conversation of climate control continues to be a hot topic, it seems, like many of the issues contributing to climate change, to be a Western problem. Museums in countries like the Philippines, Nigeria, and Benin have successfully maintained their collections without the use of HVAC systems. At the National Museum of Lagos, conservator Ogechukwu Okpalanozie describes the collection care approach of her museum: the collections are made from materials found in Nigeria, so the materials are used to the natural environment (Ki Culture 2022). It could be argued that the most damage incurred on materials was taking them out of their natural or historical environments and putting them in climate-controlled spaces in the first place.

Even many museums in Europe have collections in non-climate-controlled spaces. The Vatican Museums have the paintings and tapestry displays climate controlled while the sculpture galleries can be found to have open windows during the summer. The Municipal Museum of Ústí nad Labem in Czechia has no climate control because, according to their staff, their collection simply doesn't need it (personal correspondence, 26 August 2022).

It may be not a question of why we are not making the change, but actually why we implemented these regulations in the first place. Is this another example of academic colonialism? British professionals make a claim for collections management for their own context and then did everyone jump on board because Thomson and Plenderith's numbers came from the National Gallery in London? (Thomson 1978). And now, if a museum starts with strict standards, would it be in poor form or look unprofessional to relax standards? Perhaps, it is a matter of semantics. "Relax" as a word has an unprofessional connotation, which would deter museums from changing for fear of losing their reputation. What if these changes were updates? Would this drive adaptation? Additionally, smaller museums fear losing out on loans from larger museums and consequently cite loan agreements as deterrents for changing. "Adhering to tight parameters in loans prevents institutions from moving toward more sustainable approaches" (Boersma et al. 2018: 12).

This is again an example of the inaccessibility of culture. One may argue that loans are intended to make objects/artworks available to audiences who otherwise might never have a chance to travel to visit the object in person. However, these loans only tend to serve the tourists from blockbuster museums – the "wow" factor as museum specialist George Abungu says (personal correspondence, 16 August 2022). In this case, is it not to serve underserved communities, but only to increase ticket sales or press coverage?

While there are many guidelines (ASHRAE, AICCM, Bizot), there are no universally accepted international standards. However, in an attempt to follow "best" practice, many confuse guidelines for standards. A survey conducted in 2022 demonstrated just how disparate conditions are at organisations and the reasons behind using various guidelines (Ki Culture 2022). Joel Taylor, in a presentation to the Ki Futures program in 2023, compares the desire for a universal standard to the existence of a unicorn, "a good standard is a bit like a unicorn . . . it's almost a mythical thing . . . this perfect guidance for people" . . . which doesn't exist (Taylor 2023). While people search for these magic numbers, they turn to published guidelines and start calling them standards. And when one set of standards or guidelines are implemented, changing them looks unprofessional, so museums stick with what they have.

Upgrading climate control

This idea of following best practice can be dangerous because what is "best" practice for one may not be best for another, and this is particularly true for collections. Making assumptions or basing decisions on the work of others without applying individual context, knowledge, and critical thinking skills is never "best."

As described earlier, this is how we ended up applying strict climate conditions internationally. Additionally, assigning labels to various types of practice can make people believe it's "best" when in fact it's arbitrary. The ASHRAE guidelines, for example, range from AA to D. Naturally, the human mind interprets these as "best" to "worst," assuming that AA would be ideal or "best" conditions for collections and that D is the worst, and maybe missing the footnote that what is most appropriate actually depends on the type of building and the type of collection, and that any decision should take into consideration historic conditions. This natural condition of human nature to ascribe "best" and the human desire to do the "best" can inhibit change.

The climate control conversation is just one of many areas in which there has been discussion, but little movement. This reluctance to change has been based on fear, misunderstanding, and the reliance on peers to set the standards of "best" practice. Museum professionals, in particular, have an innate sense of professionalism. Once certain standards are set or levels of practice are reached, change to less strict standards can feel like going backwards or even failure. As cited previously, this idea of professionalism inhibits change.

Collecting

When we look at collections management with a sustainability lens, it makes little sense. Museums continue to accession without adequate deaccession strategies and build larger and larger storage facilities to house their collections (Merriman 2008). All while 90% of museum collections are never even seen by the public. Museums claim that preserving their collections is for the public benefit, often relying on public funding, then claim that there isn't enough funding for environmentally friendly changes or socially conscious practices, directly betraying their societal responsibility (ICCROM 2016; BBC 2011). The economic models of the sector are broken. The environmental models weren't there in the first place. And the social aspects are exclusive and outdated.

The preciousness of collections and objects prohibits institutions from selling off items from their collections (even when to pay salaries or for operations during a global pandemic) or from deaccessioning objects that have little or no value – historical, artistic, or otherwise. This conservative approach is the fundamental purpose of a museum to house and protect objects for the public posterity. But given the current climate and crises in the world, this purpose may need to be revisited to see if this construct is still relevant. As museums continue to add to their collections – inheriting and buying new objects, refusing to let go of the old or the contested – they build new and larger storage facilities, the embodied carbon and energy consumption of which results in an exponential increase in the carbon footprint of the sector. And if the vast majority of those collections remain in storage, not to be seen or enjoyed, then who are we preserving these treasures for anyway (Keene 2008: 72)? And at what price?

Conservation

The largest carbon footprint in conservation has already been addressed earlier in the discussion on climate control. But conservation is more than just temperature and relative humidity – and has a bigger environmental impact. Most conservators come to environmental sustainability because of their day-to-day experience with materials – throwing them away or contact with dangerous, toxic substances.

Conservation is about the preservation of our past for our future; it is the act of protecting tangible (or intangible) heritage. Conservation ethics traditionally focus on the object – how can we preserve something as it is or as it was? – rather than focusing on bigger picture impact. It is ironic that a profession so focused on long-term preservation can forget about the preservation of the planet.

While the remainder of this section will analyse the environmental impacts of the conservation field, sustainability is completely interconnected, and it is impossible to talk about the climate crisis without talking about social justice without discussing Indigenous rights. Therefore, it is only appropriate to take a moment to address the elephant in the room regarding ethnographic collections and repatriation. In some cultures, objects are not seen as objects and instead seen as living beings with a soul. In an interview, Nicholas Galanin – Yéil Ya-Tseen, Tlingit/Unangax̂, a contemporary artist –explains his work expressing this,

> As an Indigenous person, our historical inheritance, objects, even ancestral bones are often held in museums. So, this work, *Escape*, was showing a way out of these prisons, which is what museums are essentially. And the way out is returning the objects.
>
> (Simon 2020)

Conservation focused solely on materiality is outdated. It is no longer valid to claim that material value and preservation take precedence over cultural value or societal understanding. It is the author's hope that she will live to see the day the Parthenon Marbles are returned to Athens.

The methodology and approach to conservation techniques vary globally, and ethical aspects of heritage are becoming more widely considered. But the conservation of a material or an object, no matter what ethical or moral standards are respected or what social aspects are considered, is moot if there is no future to preserve these materials and stories for. The idea that we are preserving our past while harming our planet and ourselves in the process just shows how drastically change is needed. Conservation practice today is just starting to take into account ecological impact, and this is seen particularly in discussions regarding materials, packing and shipping, lighting and climate control. However, some holistic questions and underlying systemic problems are skimmed over in these discussions. A deeper look into the fundamental problems of the sector reveal solutions but requires holistic approaches.

Conservators are inherently conservative but are also taught to be risk averse. This can present a challenge to making changes for the climate. But this also affects materials used in treatments. Practice today requires conservation materials to undergo decades of research and artificial ageing tests to prove their effectiveness before use. With this in mind, it is amazing to think back to how these standards were not applicable when it came to the plastics boom in the 1950s. All of a sudden if it was synthetic (polymer-based acrylic resins or man-made chemicals) it had to be better. Conservators applied cellulose nitrate or polyester resin, which today have degraded poorly and now pose issues for removal. This lesson was hard learnt, and conservators are more cautious to apply materials, requiring years of research to ensure no detrimental long-term effects. This has inhibited the adoption of sustainable alternatives and the re-evaluation of traditional techniques which survived for hundreds of years (Vatteroni 2020). The hesitation to use naturally derived products because of potential long-term effects or contamination seems to be founded in a societal antipathy for dirt.

We have been taught to fear germs, bacteria, and dirt (Germ Theory was developed around the turn of the 20th century) and taught to celebrate clean – anything antiseptic, antibacterial, "killing 99.9% of germs" – was a good thing. In the 1950s, marketing campaigns made germs the bad guy, and since then the tendency is that cleaner is better and removing all bacteria and dirt is healthier (Tomes 2000). However, removing essential bacteria from our guts has actually had detrimental effects to our health., and we have seen a decrease in the human immune systems – but of course we just increase our dependency on pharmaceuticals (González Olma et al. 2021). Conservation is very much the same. We made this giant leap of faith to science-based and synthetic, growing fearful of the very Earth that we come from. This separation of humans and nature is a root cause of the climate crisis and the various injustices that our society is experiencing today. Jane Goodall says that to solve the climate crisis, we need to reconnect with nature (2022). Humans too often forget that we are part of a larger ecosystem. That we are not above nature or separate, but that we are part of this Earth and connected to everything on it.

The fundamental question at the end of the day is why do we practise conservation? And how? The answer is simple: for people. For us. Conservation and restoration of cultural heritage are an inherently selfish undertaking. Not in a bad way. Sustainability is also primarily selfish. We want to preserve this planet as inhabitable so that humans can continue to live here. The Earth will be fine without humans, but do we survive without Earth? What we are trying to accomplish is the preservation and continuation of our species and our legacy – so that we may remember where we came from and who we are. But this practice has become lost in the same way that our species has.

Fundamental questions are posed. Are we responsible conservators if we are destroying the planet in the process of preserving material objects? Are we preserving culture if we don't communicate with the community where materials were created? Or consider ourselves and our own health and sanity?

Despite what we inherited, conservators can, and must, be the solution. We are equipped with a unique skill set and need to start applying critical thinking skills to conservation approaches that are more environmentally and socially, not to mention financially, sustainable. This new approach and value system also need to be integrated into education systems. Trends towards sustainable education and integration into conservation programmes have begun, but need to be streamlined, empowering not only students but also professors and lecturers to feel comfortable and confident in understanding and teaching sustainability. Examples of integrating sustainability into conservation training can be seen at the UCLA-Getty Program, Queen's University in Canada, the University of Glasgow, Metropolia in Finland, and by the four Swiss conservation programmes (examples are taken from personal correspondence, Graham 2022; Herriges 2022, and the author's direct involvement in these programmes).

The future

In order to ensure the future of the heritage sector, change needs to be made. Many museums around the world are already demonstrating how to be relevant: how to effectively and appropriately serve their communities and audiences and how to be environmentally responsible and work with the changing seasons and local climates to ensure proper care for their collections. The Guggenheim Bilbao made an enormous statement when they increased the ranges on their climate control and told lenders that they would simply omit objects from shows if the lender didn't agree with the new climate conditions (Marshall 2023). But much of the sustainability work that is going on is happening under the radar, bordering on green hushing. It is the biggest museums who are usually the slowest to change, who have traditionally set the tone for the field in terms of best practice and setting standards. However, when it comes to sustainability, many smaller institutions, with more flexibility, are taking the lead – museums like the Eco-Museum Tata Somba, which was literally built by the community for the community, the Museo Moderno in Buenos Aires, the Anchorage Museum in Alaska, the District Six Museum in South Africa, the National Museum of Lagos in Nigeria, and hundreds of community and local museums globally who are effectively serving their communities and embracing sustainability on multiple levels.

These issues, bottlenecks, obstacles, and barriers are a thing of the past. The climate crisis is here. Museum professionals, conservators, artists, curators, and institutions are taking action. In 2018, ICOM established the Working Group on Sustainability (ICOM 2018). In 2019–2020, numerous organisations around the world were established to tackle this issue, including Art+Climate Action, Art/Switch, Art to Zero, the Climate Heritage Network, Curating Tomorrow, Gallery Climate Coalition, Ki Culture, and others, joining established organizations and communities, such as Art to Acres, the Coalition of Museums for Climate Justice, Museums and Climate Change Network, and Sustainable Museums (now Environment and Culture Partners). Movements and solidarity groups such as Artists Commit, Galleries Commit, and Museums for Future demonstrated the thousands of professionals around the world who are concerned with the cultural world's impact on the climate and want to do something about it.

We are at a point when immobility will no longer be tolerated – where neutrality or excuses are no longer accepted. Sustainability is the most pressing issue, from decolonization and repatriation to reducing greenhouse gas emissions and eliminating wasteful practices, museums and galleries must get on board in order to be relevant and to survive – for us all to survive.

Conclusion

Historically, the museum world has been about collections, higher education, and knowledge. But in light of the current state of the planet and our species, every sector, everyone, everywhere, must change. While there are exceptional examples of sustainability in the field, the sector needs to find more urgency in taking up sustainable practices and making sustainability the core of every museum's mission.

The barriers and obstacles that the sector currently faces stem from inertia, short-term thinking, "professional standards," lack of transparency, financial unsustainability, and lack of agency. The museum has too long relied on safety nets, rooted in tradition, the status quo, and fear. Each individual practitioner and each individual museum must take responsibility for asking the tough questions, for reassessing our practices, and rethinking our values. Culture has the power to influence our audiences and shift the very paradigm our society, but it must do so by leading by example. Museums must become proactive in these conversations rather than reactive, or we risk complete shocks by impending societal pressures (i.e. to decolonise or repatriate) or mandates (i.e. government carbon goals).

It is time to change. As cultural professionals, we have to change not only our daily practice but also the perspectives of the world on who we are and what we do. Museums are no longer a leisure activity. Museums must become indispensable. In order to be relevant, we must be sustainable. And that will take courage.

References

Aho, I. 2012. Value-Added Business Models: Linking Professionalism and Delivery of Sustainability. *Building Research & Information*, 41(1: New Professionalism): 110–114.

AICCM. 2018. *Environmental Guidelines*. https://aiccm.org.au/conservation/environmental-guidelines/ (Accessed 28 November 2022).

Altamimi, H. and Liu, Q. 2022. Nonprofits May Need to Spend About One-Third of Their Budget on Overhead to Thrive – Contradicting a Rule of Thumb for Donors. *The Conversation*. https://theconversation.com/nonprofits-may-need-to-spend-about-one-third-of-their-budget-on-overhead-to-thrive-contradicting-a-rule-of-thumb-for-donors-188792?ref=museum-human (Accessed 28 February 2023).

Anderson, M.L. 2004. *Metrics of Success in Art Museums*. Los Angeles: The Getty Leadership Institute.

Ashley-Smith, J. 2018. Linking Climate Crisis to Climate Control. *Conservation Perspectives: The GCI Newsletter*, (Collection Environments Fall 2018): 4–9.

BBC. 2011. *London Museums Urged to Show More 'Hidden' Artefacts*. https://www.bbc.co.uk/news/uk-england-london-12214145 (Accessed 28 February 2023).

Boersma, F., Taylor, J., Dardes, K. and Lukomski, M. 2018. The Managing Collection Environments Initiative: A Holistic Approach. *Conservation Perspectives: The GCI Newsletter*, (Collection Environments Fall 2018): 10–12.

The Cambridge English Dictionary. 2022. *Greenwashing*. https://dictionary.cambridge.org/dictionary/english/greenwashing (Accessed 26 November 2022).

Charman, H. 2005. Uncovering Professionalism in the Art Museum: An Exploration of Key Characteristics of the Working Lives of Education Curators at Tate Modern. *Tate Papers*, 3(Spring). https://www.tate.org.uk/research/tate-papers/03/uncovering-professionalism-in-the-art-museum-exploration-of-key-characteristics-of-the-working-lives-of-education-curators-at-tate-modern (Accessed 7 March 2023).

Creative Carbon Scotland. 2022. *What Is Artwashing?* https://www.creativecarbonscotland.com/event/what-is-artwashing/ (Accessed 26 November 2022).

de Guichen, G. 2018. *Climate Control*. https://www.iiconservation.org/content/reflections-2018-iic-congress-turin (Accessed 28 February 2023).

DeMaria, F., Schneider, F., Sekulova, F. and Martinez-Alier, J. 2013. What Is Degrowth? From an Activist Slogan to a Social Movement. *Environmental Values*, 22: 191–215.

Desvallées, A. and Mairesse, F. (eds). 2010. *Key Concepts of Museology*. Paris: Armand Colin.

Ditmars, H. 2022. Climate Activists Pour Maple Syrup on Emily Carr Painting at the Vancouver Art Gallery. *The Art Newspaper*, 15 November 2022. https://www.theartnewspaper.com/2022/11/15/climate-activists-emily-carr-painting-maple-syrup-vancouver-art-gallery (Accessed 8 March 2023).

EnergyStar. 2023. *About EnergyStar*. United States Environmental Protection Agency and the United States Department of Energy. https://www.energystart.gov/about?s=footer (Accessed 8 March 2023).

Fan, K. 2014. *Energy Star: 'Why Not Museums?' Environmental Leadership, Action and Ethics*. Columbia. https://edblogs.columbia.edu/scppx3335-001-2014-1/2014/03/27/energy-star-why-not-museums/ (Accessed 18 February 2023).

Gayle, D. 2022. Just Stop Oil Activists Throw Soup at Van Gogh's Sunflowers. *The Guardian*, 14 October 2022. https://www.theguardian.com/environment/2022/oct/14/just-stop-oil-activists-throw-soup-at-van-goghs-sunflowers (Accessed 8 March 2023).

Gill, J. 2022. Net-zero Picasso: Museums Rethink Art Shows to Cut Climate Impact. *World Economic Forum*, 22 June 2022. https://www.weforum.org/agenda/2022/06/art-museums-cutting-carbon-emissions (Accessed 7 March 2023).

González Olma, B.M., Butler, M. and Barrientos, R. 2021. Evolution of the Human Diet and Its Impact on Gut Microbiota, Immune Responses, and Brain Health. *Nutrients*, 13(1): 196.

Goodall, J. 2022. *Change Now Conference*. Paris, France, 20–22 May 2022.

Graham, F. 2022. *Personal Correspondence*, 25 February 2022.

Halperin, J. 2017. *Here's How Much Money Museum Employees Really Make*. https://news.artnet.com/art-world/survey-museum-employees-salaries-1009073 (Accessed 1 August 2022).

Herriges, M. 2022. *Personal Correspondence*, 20 February 2022.

ICCROM. 2016. *Objects in Museum Storage in More Danger Than You Think*. https://www.iccrom.org/news/objects-museum-storage-more-danger-you-think (Accessed 18 November 2022).

ICOM. 2018. ICOM Establishes New Working Group on Sustainability. *ICOM News*, 11 September 2018. https://icom.museum/en/news/icom-establishes-new-working-group-on-sustainability/ (Accessed 10 March 2023).

ICOM. 2022. *Museum Definition*. https://icom.museum/en/resources/standards-guidelines/museum-definition/ (Accessed 26 November 2022).

ICOM-CC IIC. 2014. *Environmental Guidelines ICOM-CC and IIC Declaration*. https://www.icom-cc.org/en/environmental-guidelines-icom-cc-and-iic-declaration (Accessed 8 March 2023).
ICOM Germany. 2022. *Statement: Attacks on Artworks in Museums*. https://icom-deutschland.de/de/nachrichten/564-statement-attacks-on-artworks-in-museums.html (Accessed 8 March 2023).
Indrisek, S. 2018. The Precarious, Glamorous Lives of Independent Curators. *Art Market*, 18 February 2018. https://www.artsy.net/article/artsy-editorial-precarious-glamorous-lives-independent-curators (Accessed 8 March 2023).
Keene, S. 2008. *Collections for People: Museums' Stored Collections as a Public Resource*. London: UCL Institute of Archaeology.
Ki Culture. 2022. The International Climate Control Conference. *Recording*, 1–2 December 2022. https://www.youtube.com/@kiculture (Accessed 8 March 2023).
Lindqvist, K. 2012. Museum Finances: Challenges Beyond Economic Crises. *Museum Management and Curatorship*, 27(1): 1–15.
Marshall, A. 2023. As Energy Costs Bite, Museums Rethink a Conservation Credo. *The New York Times*, 3 February 2023. https://www.nytimes.com/2023/02/01/arts/design/museums-energy-climate-control.html (Accessed 3 February 2023).
Merriman, N. 2008. Museum Collections and Sustainability. *Cultural Trends*, 17(1): 3–21.
Michalski, S. 2007. The Ideal Climate, Risk Management, the ASHRAE Chapter, Proofed Fluctuations, and Toward a Full Risk Analysis Model. In *Contribution to the Experts' Roundtable on Sustainable Climate Management Strategies*. Tenerife, Spain: The Getty Conservation Institute.
Museums Association. 2022. *Pay in Museums*. https://www.museumsassociation.org/campaigns/workforce/pay-in-museums/# (Accessed 18 November 2022).
New York Times. 2014. After the Exhibition Finding New Uses for Displays. *The New York Times*, 20 March 2014. https://www.nytimes.com/2014/03/20/arts/artsspecial/after-the-exhibition-finding-new-uses-for-displays.html (Accessed 1 March 2023).
Plenderleith, H.J. and Philippot, P. 1960. Climatology and Conservation in Museums. *International Center for the Study of the Preservation and Restoration of Cultural Property*, 248–253.
Pogrebin, R. and Small, Z. 2022. Selling Art to Pay the Bills Divides the Nation's Museum Directors. *The New York Times*, 19 March 2021. https://www.nytimes.com/2021/03/19/arts/design/deaccession-museum-directors.html (Accessed 16 May 2022).
Richman-Abdou, K. 2018. *How Museums Evolved Over Time From Private Collections to Modern Institutions*. https://mymodernmet.com/history-of-museums/ (Accessed 28 November 2022).
Simon, L. 2020. When Will We Return What We Took From Indigenous Peoples? *Tero Magazine*, 21 November 2020. https://oneresilientearth.org/when-will-we-return-what-we-took-from-indigenous-peoples/ (Accessed 28 February 2023).
Stolle, D. 2022. *'Greenhushing' on the Rise as Companies Keep Quiet on Science-Based Targets'*. https://earth.org/greenhushing/ (Accessed 18 November 2022).
Taylor, J. 2023. *Personal Correspondence*, 21 February 2023.
Thomson, G. 1978. *The Museum Environment*. London: Butterworth-Heinemann.
Tomes, N. 2000. The Making of a Germ Panic, then and Now. *American Journal of Public Health*, 90(2): 191–198.
Vatteroni, C. 2020. Analysis on Adhesives for Stone Conservation: A Preliminary Evaluation of Potential Sustainable Alternatives. *ECR – Estudos de Conservação e Restauro – 2020*, 11(67).

Vetter, D. 2021. Oil and Coal Firms Guilty of 'Great Deception' Through Greenwashing, Say Climate Lawyers. *Forbes*, 19 April 2021. https://www.forbes.com/sites/davidrvetter/2021/04/19/oil-and-coal-firms-guilty-of-great-deception-through-greenwashing-say-climate-lawyers/ (Accessed 7 March 2023).

Weisberg, R. 2023. Museums and Their Workers Have a World Full of Problems to Deal with. *Museum Human.* https://www.museumhuman.com/links-of-the-week-february-24-2023-getting-back-to-museums/ (Accessed 28 February 2023).

9

THE 100-YEAR FUTURE

Museums and the climate and nature crisis

Maria Balshaw

The COP27 meeting in Egypt in late 2022 and the related reports from the Intergovernmental Panel on Climate Change on the impact of global warming highlight the incontrovertible damage we as humans are enacting on the planet and raise an even louder alarm that if we do not act now, much greater damage will be inevitable. As institutions of the long term, and as institutions that are fundamentally concerned with the actions, ideas and material culture of humans as they operate across time and across widely varied global contexts, museums must act and must act differently in our present, if we are to preserve conditions for the future. Without this, the heritage we are responsible for will cease to have relevance or meaning. There is still a window of time for us to steer away from the very worst of this crisis. Professor Debra Roberts, co-chair of the IPCC, said:

> *Our report clearly indicates that places where people live and work may cease to exist, that ecosystems and species that we've all grown up with and that are central to our cultures and inform our languages may disappear. So this is really a key moment. This is the decade of action, if we are going to turn things around.*[1]

For me, this brings to mind a piece of public art called *Plunge* created in 2012 by artist Michael Pinsky (Figure 9.1). As is so often the case, artists imagine the future for us and point to the uncomfortable truths that we don't want to look at.

The work encircled noteworthy buildings in London with an illuminated blue line, showing the predicted sea level in the year 3012. The public making their way around London was invited to conceive of an imaginary line between the monuments showing just how much of London would be below the water's surface. At another, the illumination suggests a protective shell proposing that we still have

DOI: 10.4324/9781003347606-10

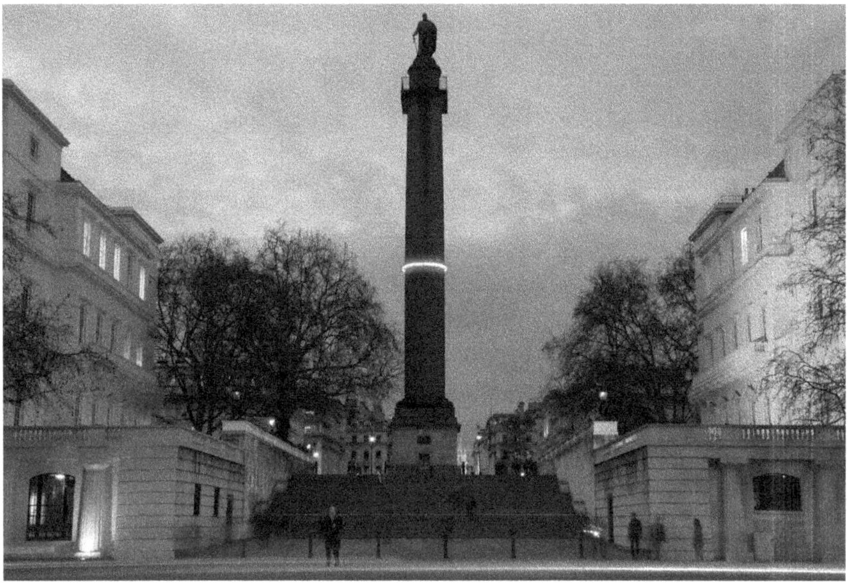

FIGURE 9.1 Plunge by Michael Pinsky. Mounted on the Duke of York Column, The Mall, London. Photo credit: Kristian Buus

chance to change this situation. The overarching message, however, is that adaption is essential if we want our low-lying capital city, in all its glory, to continue to be habitable. This isn't a prediction that is 1,000 years in the future. Within our collections at Tate are the tide marks of a previous inundation in 1928 which affected paintings in store, some of which were then turned into an artwork of their own by Cornelia Parker, in her *Room For Margins* which she used to draw attention to the climate and nature crisis she and we are living through. Colleagues in Australia have recently been dealing with the aftermath of flood waters rising to the second-floor ceiling of Lismore Regional Gallery as the Wilson River has inundated low-lying areas. Collections have been significantly damaged, but as Director Ashleigh Ralph said, the gallery losses pale in comparison with those of the community, many of whom have lost everything. This was the second 'once-in-a-century' flood in five years.

We know artists often sound the alarm about the crises we are moving toward. But as museum sustainable development leader Henry McGhie has argued (2022), we have known about this 'crisis' of nature and environment for almost a century and are yet living and acting as if it will not happen, when at least some of the artists' imaginings and the scientists' predictions will undoubtedly come to pass. As the Director of an organisation where all four buildings face the water, I need to think as if Pinsky's blue line is possible within my lifetime, and I need to act to mitigate this within a ten-year horizon. Despite the very many people who champion this issue, museums as a sector are not shifting their practice fast enough to make a difference.

The 'alteration of ecosystems' that we are now seeing across the globe is an interconnected, interdependent labyrinth of disrupting environmental crisis, threatening physical and mental human health, and the reliance of global culture, along with so many industries and disciplines, on the free flow of people. Until relatively recently, there has been a belief, both individually and collectively, that we exist somehow separately to the natural world, or at least alongside it. This separation has allowed for unbound productivity. It has enabled humans to look at nature as a sort of unlimited repository of resources. But the cost of this mentality has, of course, been vast. Museums have absolutely been party to this estrangement. The past decade has seen, thank goodness, a major shift. We have long understood that we are at the mercy of nature, but we are finally coming to see that we *are* nature, we embody it, our edges are not distinct and as cultural institutions, we need to lead the way in demonstrating that saving nature is saving ourselves: our people, our collections, our buildings, our lives. But these are easy words, and they are not always followed by distinct shifts in how we operate as museums.

The conceptual shift away from perfect preservation

What has not yet happened, in my view, is a fundamental shift in our deep thinking about how we operate our museums as organisations of and for the really long term. Ultimately, the message we are still giving out is that conservation of the objects is more important than conservation of the planet. There was a notion once that collections were being preserved for a far-off and unchanging 'posterity'. But now it's not clear what that posterity will look like, and in the meantime, the world is changing and getting hotter around us. And, furthermore, our key problems are not scientific or technological – they are political, economic and cultural and we need to fundamentally change people's habits and behaviours, including our own.

Museums have begun to share knowledge, support and advice as well as building courage towards action and spelling out how museums can be used to engage people's emotions towards being part of a larger change. There is now excellent guidance available on decarbonising buildings and switching to lower carbon operating modes. The Theatre Green book (2022), developed through collaboration across UK theatres and with support of Buro Happold consultants, the engineers leading the field in greening arts infrastructure, offers comprehensive guidance and practical support, and its buildings section has been expanded to include all Arts Buildings (Arts Green Book 2022).

Tate, along with a number of other major UK museums, is part of the Bizot Group of international lending collections, whose Green protocol (CIMAM 2014) has been in place for a decade. This states that blanket environmental parameters should be replaced by a more flexible approach, and that, where appropriate, care of collections should be achieved in a way that does not assume air conditioning or other high-energy cost solutions. The issue is that many institutions do not follow this sane and sage guidance.

Where we need to make a real change is by re-thinking our approach to collections conditions, which in turn directly influences conditions for storage, display, loans and touring. Despite warm words about relaxing parameters for collections, in practice a recent survey of the Bizot Group members showed that only 50% of the institutions were actually following the Group's own recommendations. The consequences are that we still see an assumption of the use of air conditioning or air handling in galleries and stores, and loan periods being restricted often to three months. This in turn has a significant impact on the number of exhibitions that museums feel they have to mount each year, at a time when many are trying to do two, six-month shows a year rather than four, three-month ones, for financial, people resource and environmental reasons. If we think about our public who engage with the objects in our care, they tell us that they would like longer to see things. So many exhibitions are missed because they move on too quickly. Our 'objects before people' thinking is actually undermining what museums are created to do, which is to share collections with the public.

The potential to evolve our business models and also amplify our common mission to deepen people's engagement with our collections and exhibitions also depends on these core 'conservation conditions' changing. Thinking in terms of sustainable development goals – where people, planet and prosperity are seen as the triple bottom line for understanding museum impact and strategy – requires us to make a shift. Current stipulations or habits around couriering and transport also make achievement of sustainability targets difficult. During the pandemic lockdowns, many objects moved around the world under 'virtual' supervision. There is, however, surprising resistance from many institutions to keep this very sensible mode of working as the 'default' norm for museums' object exchange.

So how should museums as a whole proceed through this crucially important decade towards 2030? Most larger museums now have sustainability strategies, drafted in the past five years, setting out their goals. Tate has a Climate Emergency Working Group, which built on the good work in place across our Estates, conservation and collections care teams and which has accelerated progress since we declared a climate and ecological emergency in July 2019. We are committed to reducing our carbon emissions by 50 per cent by 2023, and we have achieved this ahead of target, to make tangible progress towards net zero emissions by 2030. The urgency around this and the sense that we are learning as we go are reflected in the fact that there is an emerging consensus that aiming for 'net zero' is actually unhelpful. Net zero means offsetting, which is not always a sound or reliable practice and in fact gets in the way of thinking about the really radical behavioural changes that are required to get to absolute zero. Better, many feel, to aim for absolute zero and fail, rather than set a lesser net zero goal that may contribute future problems of its own.

Awareness raising and engaging the public as positive advocates for change have been particularly successful. So again, in my own organisation, Frances Morris of Tate Modern has really focused minds on how our programme and our collection can amplify issues with projects like Ackroyd and Harvey and Ben Okri's project 'On The Shore 2021' (Figure 9.2).

Yet, even in an organisation where there is passion and commitment of senior leadership time and resource, we are not moving fast enough. It is as if there is a kind of treacle in the machine, slowing down change, for fear of what the future might hold. However, if we do not accelerate on these issues, the future will be very damaging indeed. Next, I sketch out a brief checklist of how I believe this acceleration may be achieved, not only addressing practical concerns but also pointing to notions of equity and integrity that should not be neglected.

Decarbonise our buildings, our operations and our working practices

Decarbonisation is non-negotiable and is happening across all sectors of public life. Museums have a particular obligation to do this because we presume that we should still exist as institutions and collections in a hundred years' time. We need green 'invest to save' models to be core to future development of the museum landscape, making the case to government for infrastructure funds, tied to stringent carbon reduction targets. Carbon literacy should be a mandatory part of museum

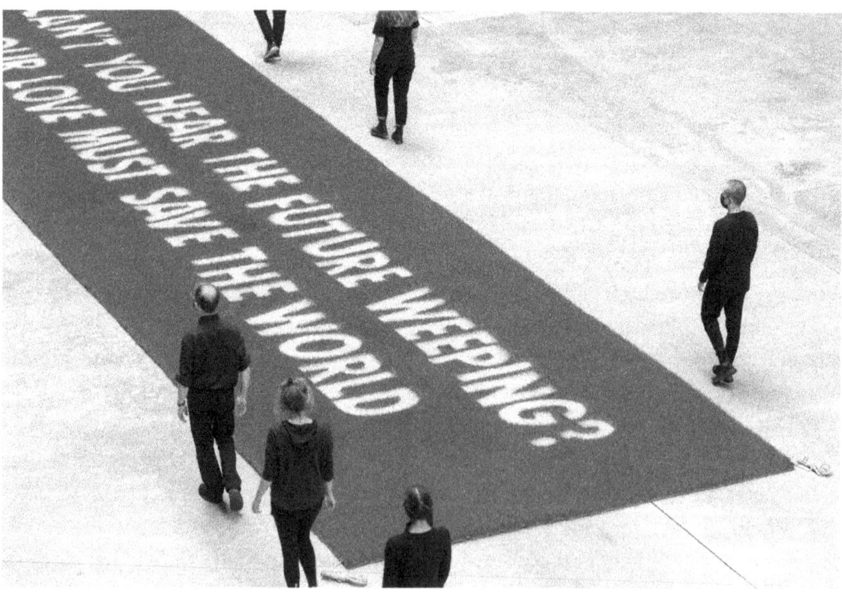

FIGURE 9.2 Ackroyd and Harvey and Ben Okri's project 'On The Shore 2021', Tate Modern, Turbine Hall. June 2021. Act 1

training so that all staff are aware of both the issues and their own roles in mitigating them. In the UK, the National Museum Directors' Council is lobbying that the next phase of government maintenance support should be directly tied to decarbonisation. Some museum directors are anxious about this, because old buildings are difficult to modify and they have many present time needs. But being hard to do should be an even bigger motivation. The international museum community should be collectively advocating for a green museum investment revolution, to protect our long-term future.

Listen to the artists

Artists are some of the most pragmatic, clear-eyed people out there. They are philosophers, forerunners and agitators. They have a way of predicting what is to come, an idea that I will come back to. And ultimately, many now have a visceral grasp of the fact that we are out of options and need to make practical changes. This practicality also does not need to exclude inspiration or beauty, so they are one of the key tools in cultivating empathy and emotion towards behavioural change. Art can act as the early alarm system, as we saw with Michael Pinsky's work. 'Artists often channel the future', says critic Jerry Saltz, 'seeing patterns before they form and putting them in their work, so that later, in hindsight, the work explodes like a time bomb' (Saltz 2016).

We see this very clearly if we look back over the past 50 years of art practice. In the 1970s and 1980s, climate change was a hazy concept that was slowly moving from niche environmental science circles into the public consciousness, with the previously mentioned IPCC forming in 1988. Movements such as 'Land Art', pioneered by Robert Smithson, and practitioners such as Gustav Metzger, or Joan Jonas, were questioning the costs of human impact on the land, the destructive impact of overproduction and overconsumption and looking to older orders of knowledge, drawn from indigenous land stewardship practices significantly ahead of clear scientific or political narratives about the need for change in human behaviours.

The need for this to be an artistic statement engaged with public, with a visible call to action, goes back many decades. For example in 1982, with the help of the Public Art Fund, Hungarian born Agnes Denes created the landmark *Wheatfield – A Confrontation*, planting two acres of wheat in downtown Manhattan. Two hundred truckloads of dirt were brought in and 285 furrows dug into the ground, by hand. For months, she and her team of assistants rolled up their sleeves and became agriculturists. Just two blocks from Wall Street, on land that was valued at $4.5 billion, the fruits of her labour grew, and by August that year, 1,000 pounds of healthy golden wheat sat serenely in its concrete surroundings. It spoke of mismanagement, waste and world hunger. Denes told the New York Times that the idea was 'an intrusion of the country into the metropolis, the world's richest real estate' – as much a critique of the economy as a protest for environmental awareness. The harvested grain eventually formed part of an exhibition entitled 'The International Art Show for the End of World Hunger' and travelled to 28 cities worldwide. The seeds were taken home by visitors and planted. Decades later, Extinction Rebellion infiltrates our cities with a very similar message.

The notion of using the natural or built environment to impart an urgent message has of course continued across the decades. In 2014, Olafur Eliasson created *Ice Watch* (Figure 9.3), for which he shipped blocks of ice to a Copenhagen square. He wanted people to have a 'direct and tangible experience of the realities of melting arctic ice'. Much like Agnes Denes, he was juxtaposing a fragment of raw nature with the artificial comforts of urban life, striving for symbols potent enough to break through into public consciousness. Eliasson repeated this project outside Tate Modern in 2018, followed the next year by an exhibition *In Real Life* which was itself conceived to be a model of how to make a show in the lowest impact way as part of the public education mission.

The exhibition had its own carbon calculator, shared with the public, and was frank about its own mode of creation and consumption. This included investment in creating circular economy product for the shop (T shirts) and a Locovore lunch menu in the café. By this point in Eliasson's own practice, and in the context of mass public engagement and activism around these issues, we can see that the artist has to walk as well as talk the issues. Perhaps the simplest but most shocking of the art included in the show were his glacier photographs, which showed all too starkly the massive loss of glacial ice over a 20-year period.

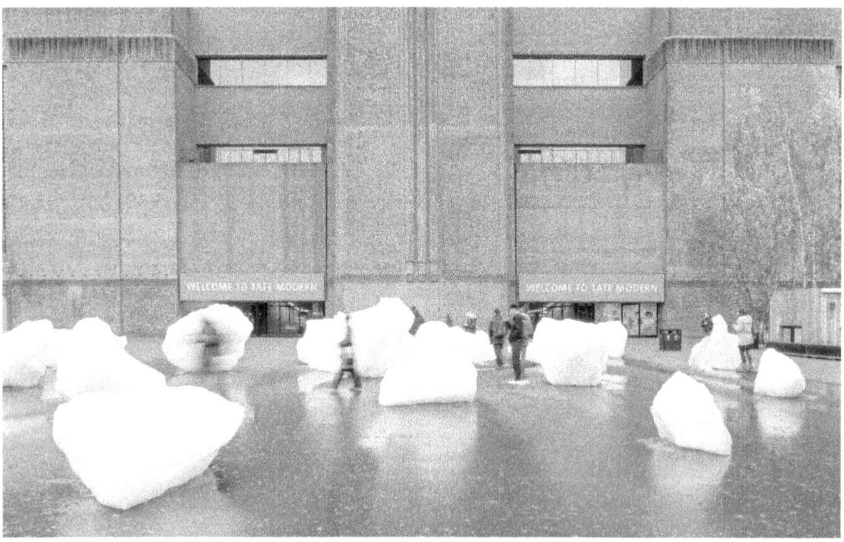

FIGURE 9.3 Hyundai commission, Turbine Hall, Tate Modern 2 October 2019 – 5 April 2020. Fons Americanus by Kara Walker, 2 October 2019 – 5 April 2020. Photo ©Tate.

It is easy to feel disheartened when looking back at how long artists (as well as scientists) have been raising these issues, to little perceptible change. But if at Tate we embrace the idea of the Turbine Hall as a megaphone, and the museum as a potential cultural convenor, we can hold on to our purpose as institutions of idea and memory that can help pave the way for changes that we all need. Ben Okri gives us useful words:

> *We have to find a new art and a new psychology to penetrate the apathy and the denial that are preventing us making the changes that are inevitable if our world is to survive. We need a new art to waken people both to the enormity of what is looming and the fact that we can still do something about it.*
>
> *This is the best and most natural home we are ever going to have. And we need to become a new people to deserve it. We are going to have to be new artists to re-dream it.*
>
> *(Okri 2021)*

Some artists are not so much making a campaigning point but rather modelling for us how adaptation can and needs to be part of our tools for a different future. Vivien Suter's work is a good example. Based in Guatemala, she responded to her own experience of climatic disaster to allow her environment to in a sense 'co-create' her work. In 2005, a severe tropical storm flooded her studio, leaving her paintings swimming in mud. She was, at first, devastated at the scale of damage. But, Suter says,

> [A]s the paintings started to dry, I saw that the natural residue had left a sharp line across the canvases, and that the paintings underneath were still visible above this line. So I saw a new potential to incorporate this, and it unified the whole work. Since then, I have also started thinking and feeling differently about painting and the conception of my work. The experience has allowed me to see my paintings as living materials that are testament to the destruction, decay and regeneration of nature.[2]

The interchanges between art, artists and science-based work are also illuminating. The charity Invisible Dust was founded by Alice Sharp in 2009. Her aim was to raise awareness of environmental issues by bringing artists and scientists together to create impactful works of art. She initially met some resistance, with scientists being sceptical about what the art world could do for them. Thankfully, there is now mainstream recognition of the role art plays with climate change. Air pollution was an early focus of the charity's mission. In 2012, artist Dryden Goodwin turned more than 1,000 drawings of his five-year-old son, inhaling and exhaling, into an animation that was projected on the roof of St Thomas' Hospital, opposite the Houses of Parliament. The piece faced commuters and tourists with a moving image showing the 'universal act of drawing breath' which will become 'intertwined with the rhythms of the city'. The work, Breathe, was designed to draw attention to the research of King's College Professor Frank Kelly, an expert on lung health, and an advisor to the government on air pollutants. Kelly's research, the Exhale programme, looks at the effect of measures such as the Congestion Charge and the Low Emission Zone on the lung health of children in East London. This marriage of art and science – a strong example of making the invisible visible – can pack an incredible punch.

Changing the conditions

This is the area where I feel there is the most need for change within the profession – and leadership and advocacy from directors of institutions like mine. We act sometimes as if there we can absolutely stop the depredations of time. As one very wise conservator said to me when I moved to be Director of the Whitworth in Manchester in 2006, we are only ever slightly slowing the rate of decay. And the perfect conditions for any object would mean keeping them in their stable storage and never showing them. So keeping museum collections is already a compromise between allowing the public to see it now and ensuring that the public in the future will still be able to enjoy it. However, I think our risk levels are currently set awry. Through our ability to use energy, technology and science to get to near-perfect conservation, we become exceptionally carbon consumptive. We need to talk more openly about making pragmatic choices that would protect the collections in our care but not so 'perfectly', prioritising instead the planetary impact and the benefit to people from learning about and enjoying the art.

Museum directors today would do well to remember the conditions of the past. I revisited the Whitworth recently, which has a particularly fine (and fragile) works on paper collection, including many works by Turner mostly gifted 125 years ago. The wonderful archive of committee papers there tell us that in the 1930s, many of these works used to travel to exhibitions at the National Gallery and Tate, wrapped in anonymous brown paper, taken by hansom cab to the station and then popped on the train, to be collected the other end by another cab.

At the Whitworth and at those London galleries, they were also displayed in unconditioned spaces. I'm not suggesting a return to quite these habits, but the fact is that the works have been passed down to us largely unscathed. And there was science behind conservation understanding then, including keeping conditions stable, not allowing too much light exposure or storing collections during WW2 in the super stable Cheshire salt mines were all evidence-based solutions of their time. We have well-evidenced research from the world-leading bodies such as the Getty Conservation Institute (e.g. Boersma et al. 2014), but we are not consistently using this data to inform our risk tolerances. We need to relax: relax the parameters for objects, focus on slowing the rate of change, but most of all we need to relax our mind set about collections. I am currently energised and heartened by the actions of conservation colleagues across the Bizot Group who are, with the support of key Directors including ones from Washington DC, Hong Kong, Sydney, Toronto and Los Angeles, supporting each other to pull together the global evidence base to give the confidence to the other 50% of Bizot leaders and the world museum fraternity, so we can adopt evidence-based judicious long-term care approaches. This is anyway most likely the best way to really care for our collections in the long term: most conservators will tell you that the greatest danger to an object is from the abrupt failure of conditioned environments, something that will become a common occurrence in a world of more extreme weather events and energy scarcity.

Finally the 'perfect conditions' we set for our objects effectively exclude many countries and cultures from the Global South from equity and exchange across the global museum ecology, as they will never be achievable there. Perhaps the worst argument I have ever had as a museum professional was with a then-much-more-senior-than-me national museum director who maintained that a Ghanaian Kente cloth could not be lent to Ghana because they would not be able to demonstrate stable conditions – which that piece of cloth did not even need. I want to think and hope that the debate has shifted, but I fear that old habits and presumptions still cling on and we need to challenge ourselves to be bolder and braver.

Move art differently

Perhaps the most challenging issue facing the art world is how to move work around the globe in a sustainable fashion. The industry as a whole is carbon intensive, with work shipped from country to country along with its couriers and audiences following in pursuit. At Tate, we knew we had to re-consider this approach. It turned out that COVID, as it did in a myriad of ways, provided an opportunity for such change. Faced with international travel restrictions, we realised that we were going to have to trial a virtual-only courier approach to ensure that we could deliver on international contractual engagements, sometimes in parts of the world that were not locked down and closer to home to continue to share our collections and deliver our exhibition programme across the four Tate galleries. This was entirely new territory for us. We were used to having a courier, often a conservator, with the artwork from the moment it left our stores to the moment it was hung, many thousands of miles away.

It was a little nerve-wracking to begin with. But, in fact, this temporary arrangement allowed us to gain significant experience in lending and borrowing securely, without the safety blanket of a physical courier. Tate Modern's *Warhol* exhibition, full of high-value loan objects, was successfully deinstalled and transported to Museum Ludwig, Cologne, solely using digital communication and trackers. Similarly, *Constable: A History of His Affections in England* was installed at Mitsubishi Ichigokan, Tokyo, with Tate colleagues joining their counterparts by Zoom to oversee the handling and installation of 60 works from Tate's collection.

In January 2021, having gained confidence from working in this way, Tate confirmed that it would move to a virtual-first courier policy. To date, the policy has already seen a significant reduction in courier travel and the associated carbon emissions. We have also since signed a Memorandum of Understanding with the Museum of Modern Art in New York to reciprocally agree that a virtual-first approach to couriers will be adopted for loans between our two institutions. We are hoping that similar arrangements will soon be in place with other institutions.

Curate sustainably

There are a multitude of considerations for curators and registrars when planning for more sustainable exhibitions. From thinking locally and ensuring that works from far away lenders are entirely justified, to curating exhibitions from just one lender rather than a multitude, from exploring options of train and ship transportation rather than air and accepting the ensuing delay this has in sending and receiving works, these approaches need to be embedded as part of exhibition planning from the outset. At Tate there has been some pioneering work, where the presumption that we should strive to minimise carbon consumption has led to new ways of making and sharing exhibitions. So the Zanele Muholi exhibition saw the works being printed and framed in the UK and then travelling in compact transport to their European venues. The fact that the artworks and the artist could not travel pushed us to test this. It led to no diminution of the experience for the visitor and was financially as well as environmentally sustainable.

Materials associated with exhibitions are also up for challenge and change, from moving to fixed exhibition layouts so that we do less build, to recycling materials, so the scaffold boards that helped create the raised 'studio-like' plinth in our recent Rodin show came from and went back to the building trade. Perhaps the most beautiful example is Kara Walker's Fons Americanus (Figure 9.4).

The monumental sculpture, a critical take on imperial public statuary, was fabricated with assistance from Millimetre, an innovative eco-design and fabrication company based in Brighton. It is made from sustainable cork, then coated in Jesmonite to make it waterproof. The sculpture was designed to be zero waste, as all the material was recycled into the manufacturing process. It not only gave us a zero waste sculpture but also was practically much easier to transport and install, and meant we did not need to reinforce the floor!

Collaborate

Collaborations between museums to allow more sustainable working modes must become normal practice, from the start to the end of the process. Tate Modern's *Surrealism Beyond Borders* was conceived in collaboration with The

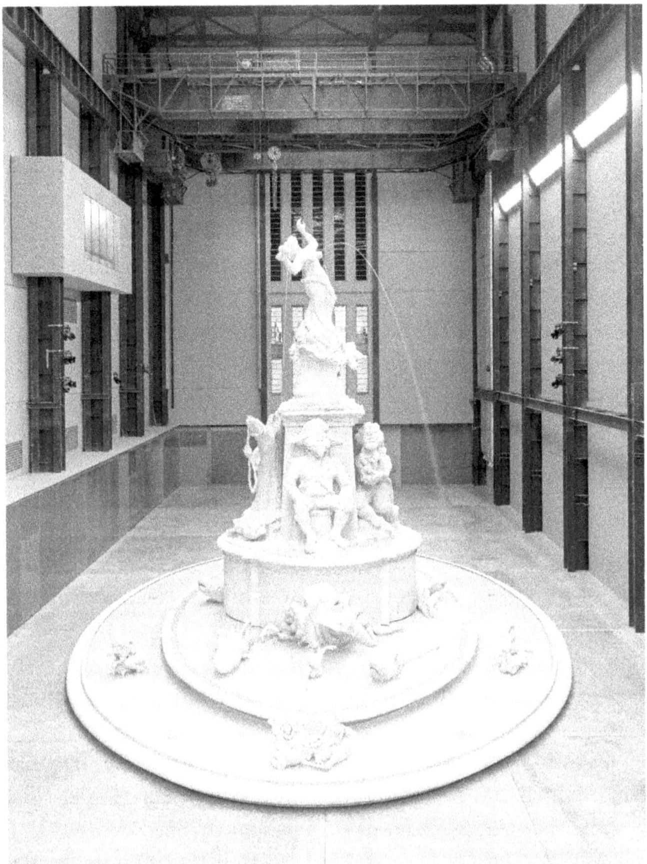

FIGURE 9.4 Olafur Eliasson and Minik Rosing Ice Watch, 2014, outside Tate Modern, 2018. Photo ©Tate.

Metropolitan Museum, New York, where it opened in October 2021, even though the project was initiated by research from a Tate curator. Teams across both museums worked closely to review the selection. The central goal was to ensure that we could continue to present a transnational exhibition, demonstrating the global breadth of surrealism, while reducing the climate impact of delivering a show of this ambition. The initial 'wish-list' included 390 loans from approximately 90 locations, but through collaborative discourse, this was reduced to 222 loans from 22 locations, with an emphasis on artworks from the collections at The Metropolitan and Tate. Active collaborative thinking across institutions can allow us to still make extraordinary international exhibitions with a much smaller carbon footprint.

Do not enable (fossil fuel sponsorship)

Museums' involvement with oil and gas companies and other environmentally damaging organisations speaks volumes to members of the public. If civic bodies appear ambivalent towards the exploitation of the world's natural resources, accept sponsorship from companies whose fortunes are made through oil and gas or are not seen to be reducing their own carbon consumption, we can expect challenge from our visitors. This seems to me reasonable and for institutions like the British Museum and the Science Museum, the level of protest and complaint around green- or art-washing is often debilitating and overshadows the work they are doing. These days this goes far beyond activist moments: the general public we welcome every day do not wish to see such ethical dissonance in museums they care about.

Making arguments with the public

Moths to a Flame, a Devon-based arts collective, invited people from around the world to hand make moths from recycled materials and to record whispered messages of hope and encouragement around the climate crisis. The moths, of which there were eventually 20,000 from participants from Devon to Australia, formed an installation, along with the soundscape of whispered messages, at the Glasgow Botanic Garden during COP26 in 2021. Organiser Chloe Uden said

> This is an important moment in time for a cultural response. In a state of emergency, our thinking can become narrowed and limited. This is an opportunity for us to show that culture can not only be part of the *response* to the climate emergency, but also part of the potential *solution*. Creative thinking can help us to imagine things that haven't happened before, to create a new future.[3]

At the Horniman Museum, the temporary replacement of jellyfish and other animals in its aquarium with the plastic pollution that blights their natural habitat was a conscious attempt to mobilise families and children as citizen advocates. The powerful letter from a five-year-old to his local MP was only one of many examples of museum visitors acting on behalf of the planet.[4]

Clare Matterson, then Director of Learning and Public Engagement at the Natural History Museum (now Director of the Royal Horticultural Society), worked with young people to develop a consultation and then a campaign for COP26 in Glasgow. The message from this younger generation was clear: 'Don't mince your words! Tell us it like it is'.[5] Museums are already trusted authorities, and our younger visitors are asking us to use that trusted expert role on behalf of them and all our futures. This 'activism' is not at odds with the museums' responsibility to share knowledge equitably and dispassionately; it goes to our core purpose – to expand and share learning for the public good.

Be tough, but real

It is important to note that none of this is straightforward. While all the points in this mini manifesto would ideally be adopted into museum practice, the basic needs of arts institutions, which, as I have said, have significant carbon footprints attached, should not be dismissed. Tate is a global organisation, and, as a result, there has to be a certain amount of travel and movement for us to fulfil our remit. The alternative to this would move us towards introspection, parochialism and lack of connection to global partnerships and collaborations that can also build strength to make significant change. Museums' sustainability strategies must be just that – sustainable – and realistic. This requires determined granular work to measure and understand environmental impact and acting accordingly.

The Tyndall Centre for Climate Change Research at the University of Manchester undertook a long-term action research programme with the band Massive Attack who wanted to address the climate cost of a music tour but still wanted to be able to play in many cities (Jones et al. 2021). The detailed work many museums are now doing to have costed 'absolute zero' plans for buildings and operations is tough, lengthy and unglamorous, but necessary if we are to change for the long term.

Encourage optimism

Tate's statutory mission is to promote the understanding and enjoyment of art. This seems quite mild and perhaps unrelated to our desire to take a strong public stand on addressing the climate and nature emergency. On the contrary, I would argue that it gets to the heart of what museums should do in the face of this crisis. We must share useful knowledge and thinking, whether this is through the insight of an artist or through offering well-argued and publicly accessible scientific knowledge. We need to encourage people to hold on to the optimism that they can affect change: the enjoyment bit.

At an extremely memorable public debate at Tate Modern between Olafur Eliasson, former Irish President Mary Robinson (now one of the global Elders[6]), and one of the leaders of Extinction Rebellion (XR), the debate got feisty as Robinson suggested XR's mode of activism tended towards alarm and doom. This, she argued, was not going to help build the energy for the long journey of adaptation and culture change we all need to pursue as citizens of a planet we want to protect. I sit between their two poles. Sometimes, as with Michael Pinsky's work, we need to look squarely at the catastrophe. Our museums and visitors need also to keep our collective courage going and enjoy the creative challenge of change, so that we can shape a long-term future where all our objects can still do their powerful cultural work.

Notes

1 https://www.bbc.co.uk/news/science-environment-60525591
2 https://www.tate.org.uk/tate-etc/issue-48-spring-2020/wild-art

3 https://gilliantaylorpr.com/2021/09/08/20000-moths-to-form-creative-call-for-change-at-cop26-in-glasgow/
4 https://www.horniman.ac.uk/wp-content/uploads/2020/03/6.-beatplasticpollution-final.pdf
5 Shared during panel discussion at conference 'Museums and Galleries Responding to the Climate and Ecological Crisis, Manchester 7–8 March 2022. https://www.nationalmuseums.org.uk/climate-crisis/nmdc-environment-conference-2022/
6 https://theelders.org/

References

Arts Green Book. 2022. https://artsgreenbook.com/sustainablebuildings/

Boersma, F., Dardes, K. and Druzik, J. 2014. Evolving Perspectives on the Museum Environment. *Conservation Perspectives: Collection Environments*, Fall 2014. https://www.getty.edu/conservation/publications_resources/newsletters/29_2/evolving_perspectives.html

CIMAM. 2014. *Bizot Green Protocol*. https://www.cimam.org/sustainability-and-ecology-museum-practice/bizot-green-protocol/

Jones, C., McLachlan, C. and Mander, S. 2021. *Super-Low Carbon Live Music: A Roadmap for the UK Live Music Sector to Play Its Part in Tackling the Climate Crisis*. Manchester: Tyndall Centre. https://documents.manchester.ac.uk/display.aspx?DocID=56701

McGhie, H. 2022. *Climate Action is Not Just for COP. Keynote Presentation at Conference 'Museums and Galleries Responding to the Climate and Ecological Crisis' held at Whitworth Art Gallery by National Museum Directors' Council*, 7 March 2022.

Okri, B. 2021. Artists Must Confront the Climate Crisis – We Must Write as if These Are the Last Days. *The Guardian*, 12 November 2021.

Saltz, J. 2016. Andreas Gursky Predicted the Future – and Present. *Vulture*, 6 December 2016. https://www.vulture.com/2016/12/andreas-gursky-predicted-the-futureand-present.html

Theatre Green Book. 2022. https://theatregreenbook.com/

10

CULTURE'S CONTRIBUTIONS TO THE CLIMATE CHALLENGE – A BRIEF ACCOUNT OF JULIE'S BICYCLE

Alison Tickell

Julie's Bicycle (JB) was founded in 2007 just before the UK became the first country in the world to enshrine carbon reduction targets into law. It was set up by the UK music industry to develop industry-led, practical responses to climate change. Within three years, the remit had expanded to performing and visual arts, including museums. JB has attempted to balance environmental 'hardware' – buildings, materials and events common to many cultural activities – with cultural software – sectors, people, networks and their specific vernaculars. This chapter takes a broadly chronological approach to the evolution of JB which, while including museums, illuminates our generic goal to shift the cultural system as a whole and fulfil our original vision of culture powering action on climate change. Our mission was to research and co-create tools, expertise and community to realise this goal. The question was how, and this chapter provides a methodological perspective which has underpinned all of our work. Needless to say, this has been the collective endeavour of thousands of colleagues and friends, whose contributions are incalculable and ongoing.

JB began with a question: what was the best way that culture could catalyse change? This question has deepened and nuanced over time, as the climate crisis reveals itself as a symptom of a culture in crisis, the values that underpin the global economy in urgent need of reform. The initial hunch was sound: that the climate crisis calls for a fully embodied response through deeds *and* words. Aside from the obvious merit of walking the walk, this was an action-research methodology intended to translate and apply the climate crisis into the cultural sector and generate the creative dynamism so lacking in climate communications. The context in which art and curation occur matters. A building, tour, production or exhibition that materially demonstrates a commitment to climate action in its operational choices – the materials, energy, and behaviours back stage as well as front – is a safer space

DOI: 10.4324/9781003347606-11

for artists and culture to champion values of care and environmental stewardship. We are all subject to our environments, conform to norms and adapt to new contexts. How we work is the DNA of why we work.

When JB started it was clear then, as it still is, that applying climate action operationally is a strategy of least risk on all sorts of levels; it serves artists and creativity as well as audiences and society as a whole. Getting this right creates the conditions for inspirational content to thrive, so this approach had potentially the most impact and possibilities for transformational change.

And arts and culture profoundly matter. They reflect our values, our habits, our unconscious and conscious biases. Choices that cultural professionals make matter, not only because they increase or decrease our carbon footprints, important though that is, but also because they reveal attitudes and conceptions of worth, what is considered consequential and what we do or don't care for.

While this is obvious now, in 2007, for most people in the cultural community, including music, it was not the case. Ennui and aversion to change, reverence of art for art's sake, the complexities of rethinking operations at scale and the perception that the environmental impacts from culture were insignificant (especially in comparison to other industrial sectors) convinced many that action was best delivered in the form of a hit record.

The perception of climate change as a peripheral issue of little concern, best bundled up into a campaign, was pervasive, and cultural policy, so vital for setting sector priorities, narratives and investment, was deeply inadequate – and it still is. While baffling in 2023, then it was hardly surprising. The politics and subsequent policy responses to climate change only took off in the 1980s and had no trickle-down into cultural infrastructure. The big moments – Brundtland Commission of 1987, the establishment in 1988 of the IPCC, the Kyoto Protocol in 1997 (ratified 2005) and the subsequent decade of fossil-fuelled geopolitics concluding in a disastrous Climate Change COP15 in 2009 – felt distant and irrelevant to cultural systems.

However, climate change was becoming increasingly important in the UK. Margaret Thatcher famously became a (albeit short-lived) champion in a speech to the UN General Assembly in 1989 and, almost two decades later, under the Labour government, the UK became the first country in the world to enshrine climate action into law. Spurred by economist Nicholas Stern's 700-page report: *The Economics of Climate Change* (Stern 2006), which made a globally compelling economic case for action beyond purely ethical arguments, the UK Climate Change Act (2008) set legally binding targets to reduce carbon dioxide emissions in the UK by at least 80% by 2050, from 1990 levels.

The Act requires the Government to set carbon budgets every five years, alongside the emissions targets, with an independent expert body, the Climate Change Committee (the CCC), to advise Government and report to Parliament on progress. Four years later, partly in response to the Act and partly in response to a determined lobby by a small group of arts organisations and individuals, Arts Council England became the first national funding body for culture which required their portfolio to report on their carbon emissions. Over a decade later, it remained the only national cultural funder to require this and to this day is the outlier.

During these years early pioneers, notably Cape Farewell, founded by David Buckland in 2001, and Tipping Point, established by Peter Gingold, were creating a cultural discourse on climate, led by artists, that changed the game. Ian McEwan, Max Eastley, Ackroyd & Harvey, Peter Clegg and Kathy Barber, Vicky Long, Ali MacGilp, William Hunt, Jody Barton, Siobhan Davies, Jarvis Cocker, and KT Tunstall began shifting perceptions and narratives in a critical way. The focus on environmental themes across culture was morphing from passive appreciation to documentation, concern, witness and, often, doom-saying. But very little was being done to the nuts and bolts of culture, the system itself.

Meanwhile in the music industry, quite distinct from the publicly funded and (largely) middle-class cultural establishment, something else was happening. David Guggenheim's widely watched docudrama, 'An Inconvenient Truth', popularised the climate crisis, just as David Attenborough did more recently. This, an extended and dramatised lecture given by the charismatic Al Gore, the almost-President of the United States, was a wakeup call. The music industry was galvanised.

Like 'An Inconvenient Truth', people had a naive faith that if everyone just understood the perils, action would follow, and the problem would be solved. JB was formally launched two days after Live Earth, a one-off climate awareness-raising event championed by Al Gore, held on 7 July 2007, echoing Live Aid a generation before. This brought together more than 150 musical acts in 12 locations around the world and was broadcast to a mass global audience through multiple media channels. This heady dynamic peculiar to the time, a combination of music industry glitz and artistic commissioning, was a manifestation of a deeply concerned cultural community that did not know what to do.

JB's focus was to apply the climate crisis specifically in relation to the activities of the music industry. Several of the first sponsors – Jazz Summers, Melvin Benn, Steve Strange – were involved in Live Earth and were promoters, venue operators and managers who were also committing, with JB, to a different kind of response. While their business acumen promoted their artists, their concerns were rooted in a deeper sense of professional responsibility: to change how they worked, rather than who they staged.

Greenwash, the idea of exaggerating or falsifying claims about environmental commitments, was rife, and Live Earth predictably experienced a backlash, its tag line morphing into 'Private Jets for Climate Change', repeating the reputational hazards already experienced by eight early sustainability pioneers such as Coldplay and Radiohead. JB was partly aiming to create the counter to greenwash to ensure that action was grounded in expertise as well as good intentions.

In 2008, JB published the first (and still only) report on the carbon footprint of the UK music industry, jointly researched with the Environmental Change Institute, University of Oxford (Bottrill et al. 2008). The research generated a new and enduring methodology for greenhouse gas emissions analysis. It identified the key impact areas (audience travel, CD packaging, building energy use) with recommendations for action. JB set up music industry advisory groups to co-create the 'how-to' resources for reducing impacts and commissioned further research (CD packaging, touring, festival travel) (Annable and Bottrill 2009; Bottrill and Liverman 2009; Campbell et al. 2022). The research findings were converted into industry campaigns and the Creative Green Tools[1]: carbon calculators for cultural and creative activities such as buildings, tours, festivals, productions that are used internationally (5,000 current organisations and artists measuring their greenhouse gas emissions).

Those early principles still guide the company:

- Data matters: to develop environmental literacy, be accountable and understand greenhouse gas emissions at a sector level over time.
- Research matters: change is hard, so doing the right thing and being able to translate this into sector-specific action can be the difference between wasted effort or targeted success.
- Collaboration matters: tools and instruments of change must be co-created with the people who will be using them.

In 2012, Arts Council England (ACE), the arm's length body charged by the government to distribute public money for culture (which distributes somewhere between £400 and £600 million to the arts and culture in any given year) embarked upon the largest programme of environmental literacy for the sector anywhere in the world and made environmental requirements a condition of funding. In response to the passing of the UK's Climate Change Act in 2008, all the National Portfolio Organisations (NPOs) funded by the Arts Council (currently some 990 receiving core support over multiple years) were asked to measure environmental impacts using the Creative Green Tools developed by Julie's Bicycle in consultation with the sector and submit to ACE a policy and action plan.

The data collection and policies formed a small part of a rich programme of research, resource-development and knowledge-sharing, the data aggregation and sector learning evaluated annually in public ACE reports.[2] JB was initially contracted to run the programme for three years; by the time we complete the current contract (2023), the programme will have run for over a decade. Supported by a thriving creative community of over 2,000 companies worldwide, from multinationals to micros, publicly funded organisations to commercial enterprises, Arts Council England environmental programme[3] supports these 990 NPOs and the ACE wider grant portfolio. Impact highlights from 2021/22 include:

- Ninety-two per cent include environmental sustainability in core business strategies

- Sixty-three per cent collaborate with other cultural organisations on environmental solutions
- Seventy-one per cent have produced or programmed work exploring environmental themes

Over time, the partnership has evolved to be more demanding. The long collaboration between Arts Council England, Julie's Bicycle and many organisations across the portfolio encouraged ACE to take a bolder approach for the 2018–2022 contract period. The data collected since 2012 is a unique resource which charts overall carbon emission profiles of England's Arts Council funded cultural sector and performance over a decade. This performance has been analysed and critiqued annually in public Annual Reports.[4] Beyond data-capturing greenhouse gas emissions, data-capturing qualitative responses to the programme form a critical part of the data set, surveying attitudinal and behavioural change: for example whether people perceived changes in their well-being as a result of taking action on climate; whether the environment made it into mission statements, board agendas; whether audiences and communities, local authorities or other collaborations were involved and, crucially, whether creative commissioning around environmental themes increased. In 2014 and then again in 2017, JB did wider stand-alone surveys to capture change over and above GHG emissions, including parts of the non-funded sector in JBs' Creative Climate Census.[5]

The Census, longitudinal tracking of the attitudes and behaviours of cultural decision-makers in response to climate change and environmental sustainability showed that between 2014 and 2017, there were significantly increasing degrees of commitment and leadership, with passionate calls for capacity-building in the arts to better respond to the climate crisis through skills, knowledge-sharing, new policy frameworks and investment underpinned by a shared vision. In 2022, JB, with Burns Owens Partnership, completed a research commission from NESTA's Policy and Evidence Centre to survey the entire UK creative sector's readiness for net zero (Campbell et al. 2022). The findings were remarkably similar to the findings of the Creative Climate Census, but writ large across a much greater geographic and creative skills set. Policy alignment across the creative industries has some way to go, but the sectors themselves are ready.

Efforts to meet the climate crisis have come again and again from the sector itself.

Almost all museums and heritage organisations consider environmental sustainability to be relevant to their organisational vision and mission (94%). Additionally, 83% or four in five think environmental sustainability will become more important over the next few years – by far the highest proportion of any respondent group (the next highest is visual arts at 69%).

Part of JB's evaluation data for Arts Council England showed that a small proportion of the national portfolio – those with large buildings including several museums – were responsible for a significant proportion of overall portfolio emissions. These institutions were also the recipients of more public funds and therefore held greater responsibility. So in 2018, JB embarked on the first programme specifically for a cultural cluster anywhere to develop credible consistent pathways to support net zero ambitions. The Spotlight programme is a cohort of 30 organisations located across 80 sites that together were responsible for around 30% of the entire national portfolio's emissions. With support from JB, they set achievable, yet ambitious, reduction targets (Environmental Impact Reduction Objectives – EIROs) in line with the goals of the Paris Agreement using Science Based Targets (SBTs).

Since the programme began, the group has reduced its electricity usage by 20% and its gas usage by 27%.[6] Electricity use has decarbonised more quickly than gas; emissions have reduced by 40% as compared to 27% in terms of gas carbon emissions. This is partly because grid electricity is being decarbonised through the addition of renewable energy sources but it is also because of the expertise and effort that organisations, together, are putting in to achieve their targets and learn from one another. Overall, the Spotlight programme has reduced its Scope 1 and Scope 2 carbon emissions by 35%, compared to the baseline year of 2018–2019 – that is in just four years (Figure 10.1).

Overall, the portfolio had shown an impressive reduction in greenhouse gas emissions while the size of the portfolio has grown (Figure 10.2).

This explicit commitment to the Paris Climate Agreement could be rolled out much more widely across culture at large, and the scope needs to include the Biodiversity Framework established at the Biodiversity COP15 in Canada in 2022. This is an easy and necessary win for cultural policy-makers.

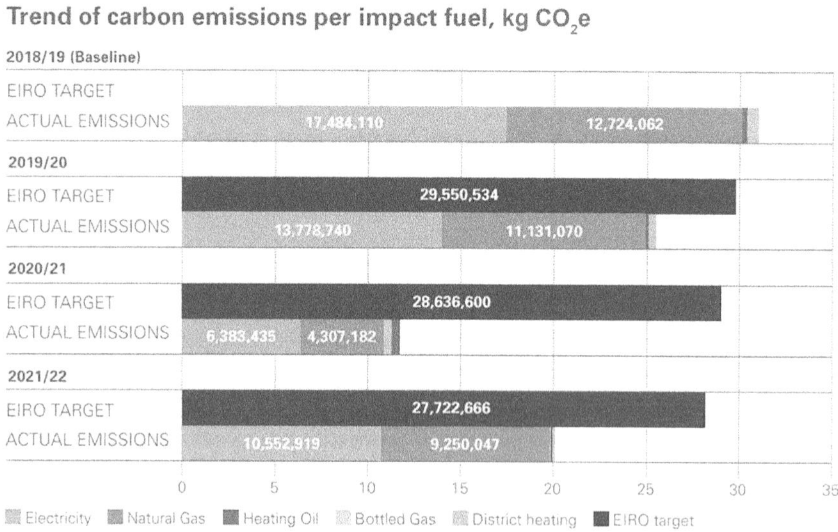

FIGURE 10.1 Reduction in carbon emissions by Spotlight members, 2018–2022 (kg CO_2e)

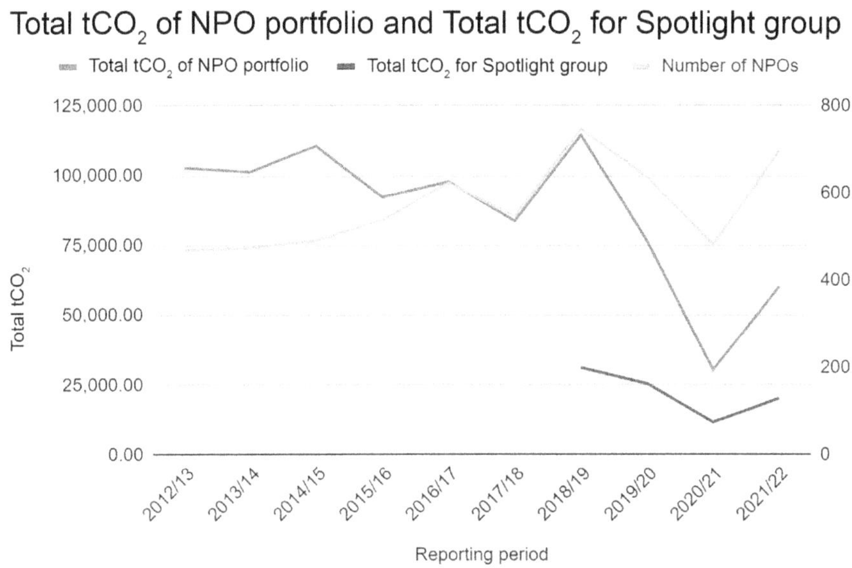

FIGURE 10.2 Total carbon emissions for NPO portfolio and Spotlight group (tonnes)

A decade of data-gathering and collective sector-based learning has generated credible and robust evidence, which makes a good case for scaling. Through all of this, JB's approach has been to demonstrate why culture and creativity deserve the same attention as science, technology and economics and to link and embed culture within broader agendas. As we address the grand challenges of our times, Arts Council England's intervention is only one, albeit effective, approach. During the last decade, both Scotland and Wales, devolved nations in the United Kingdom with their own national arts funding bodies, have committed to inspiring programmes that combine accountability and compliance with inspirational creative content.

Many of these inspirational examples come from museums, and, over the course of the decade, it is museums that have become the lightning rods for climate, nature and justice debates, where the political and ideological discussions over history and citizenship are most acute. Culture – versions of ourselves – is highly malleable, constantly sculpted into the forms and shapes of the sculptors. Museums have reflected this perhaps more than any other space designed specifically to represent the histories of Earth, excavating truths while preserving and curating culture. Significant voices in the 'contested history' debates, such as Dan Hicks (2020), Museum Detox[7] and many museums are reminding the wider creative community that we are all shaped by history and therefore have choices now to look again and reveal the deceptions, distortions and discounts written into history through curatorial and operational priorities.

Because museums have it all – material and operational infrastructures, social and environmental history holders, curators and conveners – they showcase national and global heritage attracting millions of visitors every day. These are significant institutions which have some of the most intellectually rigorous and soul-searching debates. Serving the public good, these debates are not behind closed doors either. Museums from Germany, Netherlands, the United States and United Kingdom are repatriating what have become cultural celebrities, the looted Benin bronzes. The public dilemmas and choices that museums face – most recently the protests by Just Stop Oil activists – are important fault lines between now and our future on this planet.

The imperative to change what is done, as well as what is said, was always part of their response. Throughout the last two decades, museums, like music, were engaged in their own enquiry. Leading voices including Maria Balshaw, Nick Merriman, Tony Butler, and Nicholas Serota have been at the forefront of what is now a large, vocal and significant mobilisation. For example in 2008, likely also inspired by the energy of the time, the UK's National Museums Directors Council focused on energy and environmental conditions for collections in museums and developed a set of guiding principles with the aim of minimising energy use in consultation with UK conservators, ICON and the National Trust among others. The guidelines, called the Bizot Green Protocol,[8] were accepted by the European Bizot Group of major museums in 2009 and by the main Bizot group of the largest museums in the world in 2015. But, as Maria Balshaw shows in her paper, adherence to the practicalities of the protocol by widening environmental parameters is still only patchy (Balshaw, this volume).

In 2007, the Mayor of London, Ken Livingstone, published his Climate Action Plan for London with targets and pathways that would enable the city to meet his goal of GHG emission reductions of 60% below 1990 levels by 2025. This Plan, in response to Nicholas Stern's report and the Climate Change Act, was an exciting attempt to translate the climate challenge into a particular city context and became part of a collaboration between Livingstone and the Mayor of New York Michael Bloomberg to establish the C40 city programme.[9] Justine Simons, the Mayor's Cultural Advisor, noted that culture, critical to London's identity and economy, was not included, so she set about persuading City Hall to commission a series of research reports focused on creative sectors and append them to the Climate Action Plan.

JB was commissioned to do the Music, Visual Arts and Fashion reports, while the consultancy firm The Carbon Trust covered theatre. This commission, which came shortly after the data methodology was developed by JB for the music industry, was invaluable. Using the same methodology – sector consultation alongside bespoke environmental data collection and aggregation – each guide mapped the sector's London emissions profiles and pathways to 2025 aligned with the overall Plan. In early 2010, the London Theatre Consortium, a group of 12 producing theatres, partnered with JB to develop a long-term annual stocktake of their collective emissions using JB's Creative Green framework to set targets for each following year in alignment with the Mayor's Action Plan Theatre Guide. This project yielded so much more than emissions reductions: it directly mapped to local policy for climate, stimulated a deeper understanding of the climate challenge and, over time the scope widened from building energy and travel to include productions, commissions, tours and campaigns. Cultural organisations are very good at coming together on collective commitments – initiatives that gather disparate and often small groups to coalesce into a greater whole. But doing this in service to climate action – specifically emissions reductions – was new.

At the same time, Renaissance North West, a regional hub of what was then the Museums, Libraries and Archives Council (now absorbed into Arts Council England), with the engineering firm Arup, brought out 'Museums and Art Galleries Survival Strategies',[10] setting out a five-step framework taking institutions through the process of improving energy efficiency. As a result, Renaissance North West provided further funding for two years to support a new network, Manchester Cultural Leaders Environment Forum (MCLEF, subsequently named Manchester Arts Sustainability Team, or MAST). This consortium of initially around 15 arts and cultural organisations committed to a similar collaboration on climate initiatives as the LTC. Manchester did not have a Climate Action Plan like London's until after the Paris Climate Agreement. But MCLEF's ambition was, like the LTC's, not only to set voluntary sector-aggregate greenhouse gas emission reduction targets and pathways with annual stocktakes but also to collaborate from the outset to build community and share knowledge, skills, challenges and connect to supply chains and procurement opportunities together.

Most significantly, MCLEF leadership, led by Manchester International Festival, Manchester Museum, Royal Exchange, and now the arts centre Home, included a close and iterative relationship with Manchester City Council (later extended to Greater Manchester) and the community. JB was commissioned to work with them to help achieve their vision, and the partnership has thrived ever since, strengthened by the establishment of the Manchester Climate Change Agency (MCCA) in 2015. Two significant innovations over and above MAST's considerable achievements include culture-led initiatives to achieve climate policy commitments as a result of MCCA research showing that culture, along with young people, is the most likely to engage the citizens of Manchester with climate issues; and the award of an EU grant to adapt the learning of the MAST consortium in six European cities which in turn has led to a further roll out. MAST has become an inspiration for culture-led climate action in cities not just in the UK but also more widely. This collaboration inspired many of the museums and the wider cultural sector within the Arts Council portfolio already working with JB and has been profiled internationally on many occasions. The generosity and diversity of the consortium, from major players such as the BBC and Whitworth Gallery to community-oriented and inclusion-led companies such as Home, are an outstanding example of committed, effective collaborative leadership over many years.[11]

By the time the Paris Agreement was signed and Manchester had developed bespoke city targets (one of the first cities in the world to align them directly with the 1.5°C limit on warming), JB was working with Arts Council England's NPOs, including a support programme, audits and resources for museums. Museums were rallying to find answers which benefitted the whole sector in order to make sense of the bewildering proliferation of issues and initiatives. A consortium comprising the Museum Development Network, Museums Galleries Scotland, Northern Ireland Museums Council, Welsh government and the Happy Museum, funded by Arts Council England, partnered with JB to simplify what was happening and create a framework for action. The resulting Museums Environmental Framework (2017)[12] aimed to help museums better understand their current level of environmental practice and inform their future priorities and governance. Starting with an overview of what museums were already being asked to do, the framework provided a set of 12 indicators with different levels of practice under each one, from making initial progress to leading the way. The fact that the framework was required at all was an indicator of the importance of environmental sustainability to the sector.

However, the timing of the Framework could not have been worse. Shortly afterwards the Mendoza Review, a government-commissioned 'state of the sector' report was published (Mendoza 2017). It examined the national infrastructure for museums and what was needed. While full of valuable analysis and recommendations, it made no mention of environmental sustainability nor of climate change. This critical document, which effectively summarised political and investment priorities for the sector, failed to even acknowledge the climate emergency. Published less than a year before the IPCC's Special Assessment on climate change, the report, through omission, signalled that the climate crisis was not worth bothering with. All around the country, organisations, communities, local authorities and individuals were calling for change; once again, leadership came from the sector itself. As noted earlier, JB's 2017 Creative Climate Census chapter profiling museums showed that almost all museums and heritage respondents considered environmental sustainability to be relevant to their vision and mission and that it would become more important over the next few years. Since that time, the NMDC has been actively prioritising climate and environmental action and championing the intersectionality of the climate crisis, making the links between colonisation and climate justice, nature and heritage, what is worth saving and what might go.

Policy matters and though it is changing internationally, notably in the European Commission's Creative Europe programme, there is still an urgent need to address climate and nature in national cultural policies. Much work is still to be done.

JB's research report 'D'Art 34b (Badiali et al. 2014) published by the global network of national cultural strategy and funding bodies, the International Federation of Arts Council and Culture Agencies (IFACCA), recommended that national cultural policies make explicit reference to environmental sustainability. The arts are hampered by a lack of resources and evidence to underpin cultural policy, a lack which also prevents cultural voices from contributing to solutions and influencing big climate decisions. This research took place before the Paris Climate Agreement in 2015. Even after the Paris Agreement was signed, cultural policy-makers still failed to understand how relevant and determining the climate crisis was to culture.

In the six months preceding COP26, JB revisited this question to see what progress had been made since 2014 in aligning cultural policy targets with Nationally Determined Contributions, the national targets volunteered by signatories which, collectively, need to meet the overall Paris Agreement to limit warming to 1.5 degrees Celsius (using a 1990 baseline). Simply aligning national policy and climate targets seemed like an obvious start, because the frameworks were ready to adapt to culture – specifically the Greenhouse Gas Protocol.

Furthermore, ambitious national pledges are best distributed across the economy, an omission the London Climate Action Plan had already made when it failed to include culture from the outset. With British Council support, JB combined desk-based research, a survey sent to almost 200 ministries, interviews and a series of in-depth roundtables in Turkey, Indonesia, Nigeria, the UK, Italy and Colombia to illuminate policy commitments and, perhaps, broader governance trends. The resulting report was published as 'Culture: The Missing Link' (Tickell et al. 2021) and launched at a special event at COP26 in Glasgow in 2021.

Our findings were very simple:

- Connecting culture and creativity with climate was happening all over the globe, in all creative sectors, led by deeply committed artists and activists
- In the main, this movement **was not understood, nor recognised in cultural policy**
- Aligning cultural policy to national commitments to the Paris Agreement was almost entirely absent (with the exception of Argentina)
- Yet, cultural policy champions many issues that are directly connected to climate action: inclusion (local and national environmental justice), place-making (local and national adaptation plans), democratic rights (to land, clean air, green space), enterprise and innovation (green economies) and freedom of movement

None of this was a surprise. The concept of climate justice and its relevance to culture is still not understood well enough, so the connections between inclusion, democracy, place-making and access are not being made and don't even begin to explore the opportunities for culture to develop a new, healthy and enterprising green creative economy.

Climate change action has always been, fundamentally, about justice. At the 5th IFACCA World Summit in Australia in 2011, I was asked to talk about the work of JB, and even then it was obvious that climate change is a symptom of a deep crisis in values. Climate negotiations are the battleground at the heart of a global reckoning between past, present and future, equality and justice, with the least culpable communities and economies carrying the greatest burden. This did not resonate as loudly as it should have in cultural circles in 2012, so precious time to harness the huge potential of culture has been lost.

The extraordinarily obvious connections between climate change and injustice have been articulated eloquently and insistently for several decades by people who face the daily effects of degradation and dispossession but whose voices have been consistently marginalised or ignored in political spheres. Pernicious trade agreements and enduring failure to confront the ongoing consequences of today's global economy have been too gargantuan for the United Nations, in spite of good intentions, to take on effectively. It is down to all of us to be part of this change. Julie's Bicycle has been programmatically interpreting climate justice since 2016, first developing a leadership programme for creatives from all over the world (Creative Climate Leadership run in the United Kingdom, Greece, Slovenia, the United States, Canada Benelux, Australia)[13] and explicitly since 2018 by creating a dedicated Creative Climate Justice hub[14] and accompanying programme.

Our work is just another example of the reckoning that is happening across the world as the climate and nature crises connect to the geopolitics and power dynamics that have brought us here. Although far too long in the making, countries at the most recent COP talks, COP27, formally recognised the need for finance to address the loss and damage associated with climate change and agreed to establish a fund for 'Loss and Damage'. This at last recognises the obligations of nations to finance communities and places that have been devastated by climate change and includes, for the first time, mention of cultural heritage. This would not have been possible without the intervention of the Climate Heritage Network (CHN), a global network of extraordinary focus and commitment representing cultural heritage; JB is on its Steering Group. CHN, led by Andrew Potts, had tirelessly made the links between cultural heritage and climate impacts on a local as well as global scale, showing how tangible and intangible heritage was being profoundly damaged by the climate crisis. Further, CHN had been evidencing how cultural resilience was key to managing beyond mitigation: adaptation through place-making, justice-led and democratic processes could yield a range of consensual, community-driven solutions. CHN, with its partners represented in and out of formal proceedings at COP27, established that cultural heritage matters. The cover decision for COP27, under Loss and Damage, states that:

> [T]he Sharm el-Sheikh Implementation Plan[15] '*notes with grave concern the growing gravity, scope and frequency in all regions of loss and damage associated with the adverse effects of climate change, resulting in devastating economic and non-economic losses, including forced displacement and impacts on cultural heritage, human mobility and the lives and livelihoods of local communities, and underlines the importance of an adequate and effective response to loss and damage*'.

COP27 also took the important decision to include 'tangible cultural heritage' as a theme of the new framework to enable the global goal on adaptation. That same decision also establishes traditional knowledge, knowledge of indigenous peoples and local knowledge systems as a cross-cutting consideration. Furthermore, the Sustainable Urban Resilience for the Next Generation (SURGe)[16] Initiative, one of several led by the COP27 Presidency, counts 'locally-led and culture-positive' action as one of its guiding principles, recognising that culture and heritage represent both an asset to be protected from climate impacts and a resource to strengthen communities' transformative change. The cultural policy and the climate community are, together, at a turning point on climate action, with increasing ambition to respond to the climate crisis.

Culture is vital to national economies, contributing creative skills and innovation and influencing lifestyles, tastes and consumption. But, most powerfully, art can change hearts and minds: it is intimately connected to place and to community; artists can move us to reimagine our world and inspire societies to take climate action. The creative mind put to good purpose has immense potential: to protect and preserve culture and heritage, create safe places, strengthen communities, bring joy and pride to places and people and demonstrate the future. The cultural sector – including museums – can make a crucial contribution to accelerating action.

Notes

1 https://juliesbicycle.com/our-work/creative-green/creative-green-tools/
2 https://juliesbicycle.com/?s=Arts+Council+England+Environmental+Report&x=0&y=0
3 https://juliesbicycle.com/our-work/arts-council-programme/
4 https://www.artscouncil.org.uk/our-organisation/annual-reports
5 https://juliesbicycle.com/resource/creative-climate-census/
6 See 2018/19 ACE environmental report, available at: https://juliesbicycle.com/wp-content/uploads/2022/01/Arts-Council-Environmental-Report-201819.pdf 8. The reduction targets in kilowatt hour (kWh) were calculated in line with each organisation's baseline data provided at the beginning of the programme in 2018–2019.
7 https://www.museumdetox.org/
8 https://www.nationalmuseums.org.uk/what-we-do/contributing-sector/environmental-conditions/
9 https://www.c40.org/. Global network of mayors collaborating on climate action, founded in 2005 by London Mayor Ken Livingstone, now in over 100 capital and other major cities.
10 https://museumdevelopmentnorthwest.files.wordpress.com/2012/06/museum-and-gallery-survival-strategy-guide-printable.pdf
11 https://www.g-mast.org/mast-timeline
12 https://juliesbicycle.com/wp-content/uploads/2022/01/Museums_Environmental_Framework_2017.pdf
13 https://www.creativeclimateleadership.com/
14 https://juliesbicycle.com/our-work/climate-justice/
15 https://unfccc.int/documents/624444
16 https://unhabitat.org/cop27-presidency-sustainable-urban-resilience-for-the-next-generation-surge

References

Annable, J. and Bottrill, C. 2009. *Jam-Packed: Audience Travel Emissions From Festivals.* London: Julie's Bicycle.

Badiali, C., Tickell, A. and Liverman, D. 2014. *D'Art 34b The Arts and Sustainability: An International Overview*. IFACCA. https://juliesbicycle.com/resource/ifacca-dart-report-34b/

Bottrill, C. and Liverman, D. 2009. *Impacts and Opportunities: Reducing the emissions of CD Packaging*. London: Julie's Bicycle.

Bottrill, C., Liverman, D., Lye, G. and Boycoff, M. 2008. *First Step: UK Music Industry Greenhouse Gas Emissions for 2007*. London: Julie's Bicycle.

Campbell, T., Badiali, C. and Tickell, A. 2022. *Creative Industries and the Climate Emergency*. London: NESTA. https://pec.ac.uk/research-reports/creative-industries-and-the-climate-emergency

Hicks, D. 2020. *The Brutish Museums: The Benin Bronzes, Colonial Violence and Cultural Restitution*. London: Pluto Press.

Mendoza, N. 2017. *The Mendoza Review: An Independent Review of Museums in England.* London: Department for Digital, Culture, Media and Sport.

Stern, N. 2006. *The Economics of Climate Change. The Stern Review*. Cambridge: Cambridge University Press.

Tickell, A., Schnapp, S., Badiali, C. and Tawfiq, R. 2021. *Culture: The Missing Link to Climate Action*. London: British Council. https://juliesbicycle.com/wp-content/uploads/2022/01/Climate-Connection-Report.pdf

PART 2

Case studies

11

TAKING ACTION ON CLIMATE CHANGE AND SUSTAINABILITY AT THE AUSTRALIAN MUSEUM

Zehra Ahmed and Jenny Newell

In corridors and pavilions of the global climate conference in Sharm-el-Sheik (COP27), a representative of the International Council of Museums spent a day meeting and talking with climate policy negotiators and members of organisations. Many looked puzzled when she told them the name of her organisation and asked 'what do museums have to do with climate change?'.[1]

When the world's nations signed up to the Paris Agreement, they did not only sign up to reducing emissions. They also signed up to Action for Climate Empowerment (ACE). This is action to ensure public understanding and engagement in climate change. This is a human rights issue, a duty of care by our society's institutions. Without knowledge of the current and coming impacts of climate change and without knowledge of the solutions that need to be deployed at scale, citizens cannot respond effectively. Museums have a crucial role to play. This essay demonstrates how one museum is approaching that role.

The Australian Museum: advancing action for a safer climate

The Australian Museum (AM) is Australia's first museum, built on the unceded land of the Gadigal people, first opening to the public in 1857. A museum of natural and cultural history, the 22 million collection items enable the tracing of species and cultural practices over time. Increasingly, the collections are helping the museum to talk about the future. The AM's mission is 'to ignite wonder, inspire debate and drive change'. The vision is to be 'a leading voice for the richness of life, the Earth and culture in Australia and the Pacific' (Australian Museum 2020). The museum states it is a 'dynamic source of reliable scientific information and a touchstone for informed debate . . . about the loss of biodiversity, a

DOI: 10.4324/9781003347606-13

changing climate and the search for cultural identity'. The vision is clear about the path ahead:

> We commit to transform the conversation around climate change, the environment and wildlife conservation; be a strong advocate for First Nations' culture; and continue to develop world-leading science, collections, exhibitions and education programs.
>
> *(Australian Museum 2020)*

Staff within the AM's many departments – from the researchers in the Australian Museum Research Institute to curatorial and exhibitions, Education, Marketing and Communications, Facilities, Finance and more – are all aware of the high priority the institution gives to advancing action on climate change, sustainability and wildlife conservation. The newly established Climate Solutions Centre, managed by the curator for Climate Change (Jenny Newell), is working to increase the understanding and engagement in tackling climate change. The work of the Sustainability Projects Co-ordinator (Zehra Ahmed) has reduced the institution's emissions, improved waste management and enabled certification as 'Climate Active (Carbon Neutral)' in 2020. The AM is working with communities and colleagues in the sector to help highlight strong pathways forward through a climate-changed, increasingly dangerous, terrain.

Sustainability

Becoming the first museum in Australia to achieve Climate Active under the Australian Federal government certification (otherwise known as 'carbon neutral') is the AM's way at acknowledging the enormity of the problem and the urgency that is required. Furthermore, the AM is the oldest museum in Australia. It is an important heritage symbol of Sydney, which has pivoted to adapt to the changing times. Metaphors may seem irrelevant in this space, but buildings can show how we ourselves can adapt and change.

The AM's first sustainability action plan, 2019 to 2021, was a strategic document that embedded sustainability in every department and embedded this thinking into the museum's standard business practices. To achieve the certification for being carbon neutral in 2020, the AM undertook a series of projects to reduce emissions. There is still much more work to do, but becoming carbon neutral – albeit through offsetting – has been an important way we are striving to dismiss unhelpful myths, particularly the assumption that the impacts of climate change will be felt in the future, and allows us to signal to the public that urgent action is needed to both move to a low-carbon economy and also invest and upscale the solutions that will draw down the massive amount of carbon in our atmosphere. The fact is emissions continue to rise, and this is a deeply disheartening fact. But by leading through example, and with a heritage site, we have shown that we are committed to embedding and communicating on climate change solutions. This is a space of radical hope that we all so badly need.

The AM has undertaken a carbon footprint calculation annually since 2015, and an average of the institution's yearly emissions is between 5,500 and 6,000 tons of CO_2. Knowing the carbon footprint has meant that the museum can map and identify our emissions sources and plan and implement solutions (for a good guide to undertaking this process, see McGhie et al. 2022). Having a detailed plan that identifies and offers solutions for our emissions profile takes what is understood in planning and policy as a wicked problem, to one that can be compartmentalised, delegated and ultimately archived.

In an ideal world, we would not have purchased offsets. We would have focused on reducing our emissions and profiling projects that educate the public on the importance of drawdown. But at this stage, with tied-in arrangements for energy provision, offsetting has been a requirement. It is important to note here that global emissions are continuing to rise. The solutions that exist out in the world – such as projects regenerating ecosystems – are exceptional and need resourcing now. Taking this path requires a certain amount of vulnerability, as the organisation has to acknowledge that it is still on a journey towards entirely eliminating emissions. Offsetting allows the AM to talk about climate justice and the solutions that will mitigate and draw down the excessive amounts of carbon in the atmosphere.

The AM supports a renewable energy project in India, which helps to build community resilience there. Developing countries can struggle to implement changes without the support of enabling infrastructure and resources, so supporting projects such as these give the AM the platform to call for more effort by Australian organisations for the collective race to net zero. Offsetting also permits us to elevate projects that align with our corporate strategic goals. Offsets purchased by the Aboriginal Carbon Foundation, an organisation that works with local farmers to implement First Nations-led land management, is a project that not only helps to avoid the carbon released from large-scale bushfires; it also ensures that the land being managed protects old growth-forests and supports new vegetation to grow and ecological health to be bolstered, including soil health, all of which sequesters more carbon.

The majority of the AM's emissions, 60–70%, is from the energy needed to light galleries and keep collections at the correct temperature and humidity levels. Once approval is secured to switch energy contracts to 100% renewable, the museum will have few remaining emissions.

Key projects the AM has implemented to reduce emissions were through an HVAC upgrade, which reduced our energy use by 25%, and change-over to LED lighting, front and back of house, including in collections storage. In the recent major renovation of the AM's building, a strategy for construction waste, built into the contracts for the works, meant many of the extracted materials such as timber floorboards were reused in the new exhibition spaces. More than 90% diversion from landfill for all demolition waste was achieved. Throughout the renovation, key targets within the UN's Sustainable Development Goals framework were advanced, including designing the building for accessibility by removing uneven floors and creating seamless, accessible pathways around the museum.

Throughout the site, the AM is reducing the amount of waste going to landfill. The onsite café and shop phased out single-use plastic in 2019, and in the café, plastic bottles were replaced with glass. Waste collection points gather the usual recycling streams (paper, cardboard, glass, plastic and aluminium) as well fluorescent lights, print cartridges, batteries, e-waste, mobile phones and organics. Compostable waste was added to the public waste streams in late 2020. The museum has increased its diversion to landfill onsite to 50%, and the aim is to increase this to 90% by 2027.

Making our sustainability work apparent to our visitors is an important part of advocacy and awareness-raising. In small ways, but with bigger plans, the actions the museum is taking are so far only flagged in a wall text in our permanent climate change exhibition, but our sustainability initiatives will become more visible. We have also been presenting innovative ways of dealing with waste in our temporary exhibitions, featuring innovations from materials engineers, most notably Professor Veena Sahajwalla of the SMaRT Centre, University of New South Wales. Professor Sahajwalla gives waste new life by turning it into new materials and new products. Her approach of seeing waste as an opportunity is informing how the AM talks about waste internally and the solutions it aligns with.

Future proofing

Talking about climate proofing our collections is a key part of this practice of imagining the future for museum staff. We in Sydney and more broadly in Australia are living through the effects of the climate emergency. During heat waves in Sydney, the AM has a spike in visitor numbers. Anecdotally, we know that people are making use of the museum's air conditioning to take refuge from the scorching temperature. Extreme storms have damaged the roof, causing some water leaks during heavy rains in Sydney in early 2022. The AM's scientists are bearing witness to suffering and loss of species as habitats are destroyed and more species approach extinction. The AM's conservators have been providing advice to help save collections in the line of bushfires in museums and private collections on the east coast. The next strategic piece of work the AM is about to develop is the Climate Change Risk and Adaption Planning for all aspects of the AM's operations. This plan assesses the expected extreme weather events and maps them to the museum's buildings, businesses and stakeholders. Adaptation is about trying to reduce the financial, physical and emotional fall out of expected events. It is a sobering piece of work as we map the projection for the next ten years to the next 100. How we as cultural institutions will survive the challenges of crises related to the climate emergency is uncertain. Planning, preparing and ensuring we are of service to our communities during a crisis is critical to our relevance and survival.

Exhibiting and programming for climate empowerment

The AM's Climate Solutions Centre (CSC) was announced in the mid-2022. It brings together experts in climate solutions, science and communications to support public engagement in the climate and ecological crisis. A centre for gathering ideas and creating outreach rather than a physical public space, the CSC is highly collaborative and provides a platform for diverse voices. Using many museum modes, we share powerful stories for positive futures. Underpinned by the latest social research and the AM's own extensive experience in designing engaging content, the CSC creates exhibitions and programs, including talks and film screenings, educational materials and a growing hub of digital resources and ways for people to connect and share stories. The CSC was created by a team of the AM's executive and the Curator for Climate Change, drawing on advice from local and international climate engagement experts. A strong Advisory Group has been established, bringing in guidance from leaders in a wide range of disciplines including first nations knowledge, climate science, social research, broadcasting, clean technologies, biotech and nature-based solutions.

For the museum-going public, the CSC helps to provide on-site learning experiences and opportunities for productive conversation. For peers in Sydney, the CSC helps to bring colleagues in the cultural sector together, supporting a community of interest, Sydney Cultural Institutions for Climate Action (SCICA). This group meets monthly, online or in turn at one of the members' institutions. One benefit of the group is the sharing of expertise, ideas and inspiration. Since the group came together, more staff have approached their executives with proposals for sustainability initiatives and curatorial projects. Backed by examples of what is happening at other institutions, staff have found ways of talking about the action they would like to see supported by already-tested, practical pathways and raising recognition of the changing expectations in the sector. The network is also productive for the moral support and solidarity it supplies. As many in this field know, campaigning within an organisation for climate or sustainability action can be an uneasy and sometimes lonely endeavour. Members of SCICA appreciate the camaraderie and useful connectivity with others who are navigating similar challenges.

The AM's focus on climate solutions is in response to a perceived disconnect between Australians and the methods for tackling climate change. Public engagement in climate action has been undermined in Australia particularly by the lack of awareness of the dynamics and urgency of the climate crisis, deliberate obfuscation and politicisation of the debate. The resulting polarization is accompanied by strident denialism of climate change within some communities.

The museum recognises that, as a trusted source of information, it has a highly influential voice and a clear responsibility to raise it. Cultural sector surveys, in addition to recent general public surveys the AM has commissioned (FiftyFive5 2022), demonstrate this high degree of trust in the institution and the information it provides.

The exhibitions, events and digital resources the CSC creates are supporting the AM's mission and Corporate Strategic Plan. One of the five pillars of this plan is 'advancing action on climate and sustainability'. Beyond the AM's goals, there is an overarching goal that the work of the CSC addresses: Action for Climate Empowerment (ACE), part of the Paris Agreement (United Nations Climate Change 2021). ACE centres on advancing effective education and training to empower the public to respond appropriately to climate change. Supporting public engagement was included as Article 6 of the original United Nations Framework Convention on Climate Change (UNFCCC 1992) and elaborated as Article 12 of the Paris Agreement (2015). This commitment was further strengthened in the 2021 'Glasgow Work Programme for Action on Climate Empowerment', highlighting six areas of action that can guide museums in their work on public education and engagement in climate change (UNFCCC 2021a). The full remit of ACE is to 'empower all members of society to engage in climate action, through education, training, public awareness, public participation, public access to information, and international cooperation on these issues' (UNFCCC 2021b). The Glasgow work programme reaffirmed the 'key role that a broad range of stakeholders', including government and 'educational and cultural institutions, museums [and others]. . . play in ensuring Action for Climate Empowerment' (UNFCCC 2021a).

ACE is gradually gaining recognition with more international collaborations between universities and discussions within UNFCCC fora. COP27 featured a full 'ACE and Civil Society Day' of talks, and ACE was addressed in a variety of side-events presented by cultural organisations, advocacy groups and industry groups such as the International Council of Museums (ICOM).

To support a collaborative effort among Australia's museums and galleries to galvanise and report on ACE, the AM's CSC, the National Museum of Australia, and Blue Mountains City Council have initiated a national conversation (beginning with a workshop at the Australian Museums & Galleries Association Conference in May 2023, 'Together for Action: Australian Museums and Galleries Advancing Action for Climate Empowerment'). Staff from museums and galleries around Australia are being called on to think through where their institution is placed with regard to ACE, where they want to progress, and how to get there. 'Together For Action' will establish a regular report on progress for the sector and for government, using the structure developed by Henry McGhie in his *Action on Climate Empowerment: A Guide for Galleries, Libraries, Archives and Museums* (McGhie 2022). It is expected that the process of reporting up to government, with the further aim of feeding through to national reporting at future UN conferences, will translate into more attention from museum and gallery management and, crucially, more support for taking action to increase public engagement.

Listening to our audiences

The CSC is guided by current climate segmentation and climate communications research in the development of its programs. Australian audience research shows that there is a high degree of alarm and concern around climate change. A survey of the general public (by social research group FiftyFive5, conducted in early 2022) queried people over the age of 18 who visit cultural institutions at least once a year (n = 1,513) on a range of subjects to determine their recognition and relationship to the AM. As part of this, the survey established the climate change attitudes of respondents. The survey found that 83% rated climate change as an issue of 'great deal of importance' personally. Eighty per cent felt climate change would cause harm to future generations. On a range of key indicators including the perception of the human causes of climate change, perceived risk now and in the future and the extent to which individuals can have an impact on tackling it, respondents are clustered into climate attitude segments that represent their degree of anxiety, their sense of power to take meaningful action and their optimism about the future. These segments range from 'Alarmed', 'Alert' (those who are highly concerned but disheartened), 'Concerned' to 'Cautious', right through to 'Dismissive'. The surveyed cultural-institution visitors revealed a higher proportion of concern about climate change, with 33% Alarmed (vs. 24% of the general population), 12% Alert (vs. 6%) and 17% Concerned (vs. 20%). This group had a lower proportion in the less-engaged segments than in the general population.

We also know that the AM's members (numbering almost 40,000 at the end of 2022) have very high level of alarm on the issue. An online questionnaire sent to members by email in June 2022 included questions about attitudes to climate change. There were 345 respondents, with representation across the spectrum of climate attitudes. On being asked what question they would ask an expert in environment or climate, responses revealed the depth of concern: '*How long can we keep living like this before our planet will no longer exist?*'; '*Will nature adjust?*'; '*Is it too late?*'. Also, people are wondering about the changes they'll need do make and are looking for guidance on action and something positive to hold onto. '*Where do we even start?*'; '*What are the top three things I should be doing?*'; '*How can I start adapting to the climate change we are already locked into?*'; '*How do I not lose hope in the future due to run away climate change?*' Others asked for a verification of climate change, whether it is '*Just a natural fluctuation?*', or '*Is it true?*'.

With these clear needs being expressed by the museum's audiences, the CSC has created a range of outreach to help meet them. A permanent exhibition, *Changing Climate*, opened in late 2020 in a large alcove space in an existing gallery about the Australian environment. Climate change is explained simply, the impacts that are being experienced in Australia, focusing on coral reefs, snowy mountains, rainforests and urban and rural environments, and interactives help people explore the impacts of climate and what they can do to help. Key information is relayed with a vertical digital screen stack, videos and objects mounted on three walls. A central table with an interactive map of Australia is full of entertaining animated graphics. The map is populated with a range of animals, plants, renewable technology features, EVs, farmers, surfers being chased by sharks and others, and as you touch one of these features, the characters respond. A digital 'card' pops up with questions and, when 'flipped', shows answers about climate change impacts and what people are doing to address them.

The exhibition includes the voices and perspectives of a range of people who are dealing with climate change in a direct way. Videos convey key points they want to bring to people's attention. From Dr Kim Loo, a General Practitioner in Western Sydney, urgent points about the devastating health impacts of continuing to burning fossil fuels are: the pollution, the heatwaves, the smoke from bushfires. Dr Loo underlined this point when we asked if she would be willing to donate an object to the exhibition that would help to encapsulate the impacts of climate change she is experiencing. She selected her paediatric stethoscope. It is through this, she said, that she witnesses the impacts 'on the hearts and lungs of my most vulnerable patients' (Loo 2020). Dr Anne Hoggett, marine biologist and a Director of the AM's Lizard Island Research Station on the Great Barrier Reef, donated her snorkelling mask and snorkel through which she has been watching the growing destruction of her beloved reef, which she speaks of as a canary in the global coal mine (Hoggett 2020). Two positive object stories are also included. One of the small Perspex object display boxes features a photo and quote from a farmer, Charlie Prell, in his shed. He donated some fleece from his shearing shed and the small model wind turbine that a sales representative gave him. In the quote accompanying the objects he said:

> My 28 wind turbines can each power 1,200 homes, and they mean the difference between me running a viable farm and me probably not being a farmer. Wind and solar can keep rural communities going. We have the best renewable resources in the world, why aren't we taking advantage of them? I was never an activist before, but I am now.
>
> *(Prell 2020)*

One of the photos near this quote shows a sign on Mr Prell's fence, which proclaims the group he is a member of: 'Farmers for Climate Action'. He has a steady income from the turbines through an energy-supply contact with the Australian Capital Territory, which runs on 100% renewable energy. With this reliable income stream, he keeps his own family and the families of two locals, whom he employs in his shearing shed, secure through drought and other weather extremes (Prell 2020). Phillip Julian, an arborist in Sydney, chose to donate a tag from the first tree he planted in a new greening initiative for the centre of the City of Sydney, plus one of the measuring tapes he uses to track the increase in girth of the trees in his care.

These objects – solid, personal things, similar to things we might have connection to ourselves, physical expressions of relationships to our changing environments – help us as viewers to connect to important ways of taking part in the world. They connect us to the emotional impacts of the challenges and the transition underway for people in diverse spaces within our societies. This is important. This kind of connection is a pathway to empathy. And empathy for the experience of others is a prerequisite for supportive action that will benefit – on both smaller and larger scales – the communities of beings around us.

Understanding someone else's object speaking to climate solutions can also introduce ideas for ways forward that we might not have thought of ourselves. This was the rationale driving the exhibition the CSC produced in 2021.

Spark – showcasing solutions

There were ten solutions to climate change highlighted in the temporary exhibition *Spark: Climate Innovations Tackling Climate Change* in 2021. Each of the objects selected helped to create a pathway for action for visitors. Statements in the entry areas point to the urgency of taking action to reduce emissions and the warming they bring – as 'every fraction of a degree, every year, and every choice matters' (Howden 2021). The rest of the exhibition opened up the ways that collectives of people in Australia are reducing emissions and showcased how we can all lend our weight to advancing these approaches.

The exhibition included community- and nature-based solutions, such as supporting biodiversity (including objects such as a home-made nesting box for parrots and seeds and spices made from native food species supplied by a First Nations-run nursery, IndigiGrow). A section co-produced with Firesticks Alliance Indigenous Corporation explained cultural burning – a practice of caring for Country that reduces fuel load in bushland with a gentle, 'cool' burn, a low flame applied carefully and in the cooler months, to avoid large, uncontrolled bushfires. The exhibition featured seaweed farming and regenerative agriculture (using two small models – one of an industrial, chemical-managed farm with dead-looking grey soil beneath it, another with a thriving regenerative farm with rich, living soil visible below). Objects demonstrating an exciting range of home-grown technological innovations had been sourced from research labs, small-scale manufacturers and start-ups. They included a bubbling green microalgae bioreactor; Australia's first locally produced electric motorbike; environmentally friendly concrete; designer tiles made from glass and textile waste; samples of flexible, printed solar films; experimental fluorescent solar concentrators; highly-efficient small-scale wind turbines and a domestic hydrogen battery.

After a short, intense period of curation and design, the exhibition opened in the museum's main hall in June 2021. While the exhibition was open for three weeks before the museum shut down due to COVID, *Spark* can still be visited on the AM's website as virtual tour, with the AV components included and supplying additional information about solutions that people can help to advance. The *Spark* website attracted 36,250-page views in its first 2 years and continues to draw traffic.

In Australia, there appears to be less general awareness of climate solutions than the impacts of climate change. There is less reporting of this kind of 'good news' in the media about solutions gaining momentum. Drama and disaster are still generally prioritised. The CSC's outreach provides captivating and easily-relatable content to help people find attainable, personally rewarding pathways to climate action.

A new 'Coffee & Climate' programme the CSC has devised runs monthly in the AM Members' Lounge, for a supported conversation session. The programme enables a small group to ask questions, talk about concerns in a safe space and give each person a chance to explore the type of climate action that suits them best. Informed by several local and international climate conversation models (including 'Climate 4 Change' and '@ClimateVenn'), the programme is run by the Curator for Climate Change and is designed to bolster the participants' confidence in talking about the issues, ask and answer questions and have the opportunity to consider what they can bring to dealing with the crisis. The programme has been offered to Members in the first instance to address the high levels of alarm and the many questions the group revealed when surveyed. It is working well and helping to build knowledge within the CSC around the kaleidoscope of varieties and unexpected nuances within relationships to climate change in our community.

Working with communities

The AM's Climate Solutions Centre is working to support the imagining of positive futures. It is critical, if we want to not only inform but also inspire our audiences, that museum workers provide credible evocations of the futures that citizens can create using the ideas, methodologies and technologies currently at their disposal. Humans appear to have a somewhat limited capacity for imagining and planning into the future to any great distance (Marshall 2015). There might have been few evolutionary advantages to picturing potential long-term futures in the early stages of human evolution, but there is now a distinct existential advantage in being able to picture the threat of a catastrophic future– as well as being able to apprehend the advantages of a regenerated world (Figueres and Rivett-Carnac 2020). Both scenarios have the potential to motivate action. We can inspire this visioning by providing sensorially-rich evocations of the places we can live in if we promptly manage the transition to a low-emissions, fairer, rewilded, biodiverse world. It is increasingly important to help people conjure up the world that is possible for us to create together.

The 'Future Now' exhibition – taking ACE beyond the museum's walls

In a shopping mall in a town on the outskirts of Newcastle, New South Wales (a coal port like its British namesake, the largest in the world), a retired couple and a small family peer at the details of a miniature town. Protected by a large, rectangular Perspex box set on a green base, the base carries simple pointers about the benefits of sustainability. The family and the couple point out to each other the tiny people digging in a community garden, the lush green walls enlivening the organic forms of the apartment building, the chickens in the school yard, the tiny electric vehicle chargers, the well-populated repair café and the birds they have just spotted. A school group nearby chatters excitedly and skips between a diorama of a house and productive garden and a modelled stretch of countryside, with a regenerative farm, healthy forest being cared for by First Nations practitioners of cultural burning, and a stretch of sea with a seaweed farm, wave turbine and little people restoring mangroves.

Future Now uses dioramas of ideal human–nature landscapes to show that we have what we need, now, to create the future we want. Three easily transportable 'pods' each have a landscape diorama skirted by a label rail with brief text, illustrations and a small touch screen with several videos with people who are developing and implementing key sustainability methods. QR codes take visitors to the exhibition's website, where the text is available in four languages. The AM designed the exhibition to be as accessible as possible to reach people in a wide range of demographics, locations, range of physical mobility, age range and cultural and linguistic diversity. The key messages are designed to chime well with common values that audience research has identified: affordability, clean air and clean water. *Future Now* is taking messaging about the co-benefits of living sustainably and regenerating our places beyond the museum and the metropolis. It has travelled to suburban and peri-urban areas and has been touring to regional and rural communities from early 2023. The point is made that opting for clean transport and energy and supporting nature helps to tackle the climate crisis, but most of the messages are around the benefits of making these choices to save money and create healthier, safer, flourishing places in which to live.

People seeing these ways of living have been responding positively, as the Climate Solutions Centre had hoped. It is common to hear visitors exclaim that they 'would love to live there'. One young man said he would 'fight for my own town to be like this'.

Any prompt to think about alternatives that are regenerative, supportive of community and nature, is a useful prompt. While few exhibitions can hope to be life-changing catalysts, the CSC hopes that *Future Now* will reinforce for viewers some of the new approaches and technologies that are just starting to gain visibility, or it will open up new, deeper information, or provide a trigger to look into a subject further. These functions are important for socialising any new idea or complex ideas, all contributing to the social tipping point we are aiming for. Exhibitions on climate solutions are an excellent contribution for a museum to make.

Conclusion

Advancing Action for Climate Empowerment is a responsibility that the AM takes seriously. Work is underway, finding ways to reach the disengaged, addressing audience concerns and questions, and highlighting approachable, effective solutions.

The AM is making sure it walks the talk with its work on emission reductions, stepping steadily towards net zero. It is supporting planning for a climate-changed future.

The AM is helping to create new connections with people and new conversations on climate. The museum is aiming to support comprehension of the scale and nature of the emergencies facing us. It is empowering its audiences to engage by highlighting pathways that enable us all to step up more effectively, together.

Note

1 With thanks to Katharine Turvey, Head of Capacity Building, Museums & Society, International Council of Museums (ICOM), for sharing this experience.

References

Australian Museum. 2020. Our Mission; Our Vision Statement. *Webpage*. https://australian.museum/about/organisation/ (Accessed 20 November 2022).

FiftyFive5. 2022. *Future Now – Evaluation*. Unpublished report. Australian Museum.

Figueres, C. and Rivett-Carnac, T. 2020. *The Future We Choose: Surviving the Climate Crisis*. New York: Knopf Doubleday Publishing.

Hoggett, A. 2020. *Personal Communication: Email Correspondence: Text for 'Changing Climate' Exhibition Text Panel*. Sydney: Australian Museum.

Howden, M. 2021. *Every Half Degree Matters: Prof Mark Howden IPCC 1.5 Degree C Report*. Video presentation. Website https://vimeo.com/320717385.

Loo, K. 2020. *Personal Communication: Email Correspondence: Text for 'Changing Climate' Exhibition Text Panel*. Sydney: Australian Museum.

Marshall, G. 2015. *Don't Even Think About It: Why Our Brains Are Wired to Ignore Climate Change*. London: Bloomsbury Publishing.

McGhie, H.A. 2022. *Action for Climate Empowerment, a Guide for Galleries, Libraries, Archives and Museums*. UK: Curating Tomorrow. curatingtomorrow.co.uk.

McGhie, H.A., Black, T., Indriago, G., Kapustina, C. and Chu, J. 2022. *Measuring and Reporting Greenhouse Gas Emissions: An Introduction for Museums and Cultural Institutions: Curating Tomorrow and Co2Action*. UK: Curating Tomorrow. curatingtomorrow.co.uk.

Prell, C. 2020. *Personal Communication: Email Correspondence: Text for 'Changing Climate' Exhibition Text Panel*. Sydney: Australian Museum.

UNFCCC. 1992. *United Nations Framework Convention on Climate Change*. Website http://unfccc.int/resource/docs/convkp/conveng.pdf

UNFCCC. 2021a. *Decision-/CP.26 Glasgow Work Programme on Action for Climate Empowerment*. Website https://unfccc.int/sites/default/files/resource/cop26_auv_3b_Glasgow_WP.pdf (Accessed 1 November 2022).

UNFCCC. 2021b. *What Is Action for Climate Empowerment?* Website https://unfccc.int/topics/education-youth/the-big-picture/what-is-action-for-climate-empowerment (Accessed 18 September 2022).

United Nations Climate Change. 2021. *Action for Climate Empowerment*. Website unfccc.int/ace (Accessed 15 September 2022).

12

CULTIVATING CLIMATE LEADERSHIP

Evolving institutions to address climate change and helping others to do the same

Richard Piacentini

1.75 planets

The Global Footprint Network (2022), which reports annually on the rate of worldwide resource consumption, provides one particular statistic that puts the problem of climate change and its central cause into stark relief: our current rate of consumption requires the equivalent of 1.75 Earths' worth of resources. The gap between our demand and what the Earth can provide is currently bridged by liquidating ecological resources and accumulating waste, primarily carbon dioxide in the atmosphere. Not all countries are equal in contributing to this problem; at one end of the extreme, we have developed countries that are using far more resources than is sustainable, and at the other end we have countries that are using less resources, but still are not meeting basic needs (indeed, many are aspirant to become more like these developed countries).

Our present use of natural resources is unsustainable in the long run and is proving increasingly damaging even in the short run, leading to climate change, habitat destruction, loss of biodiversity, global inequities, and climate justice issues. We may think of these phenomena as being the problem, but they are actually the symptoms of the problem. The real problem is the way we live and the lifestyles we lead. To avert catastrophe, we need to drastically change the way we live – to embrace a way that works within the limits of the Earth's resources and allows everyone, as well as other species, an equitable share.

Many of us in the museum world are working on issues related to climate change, such as conservation of habitats, preserving biodiversity and addressing climate justice issues. This is all critical and important work in the short term, but just as we often do with our own health – which is of course completely interconnected with the health of the planet – we have focused on treatment of the

DOI: 10.4324/9781003347606-14

symptoms at the expense of prevention. To bring real resolution to these issues for future generations, to stem their persistence and recurrence, we also need to start thinking about how we can address their underlying causes. We need to embrace our connection to all living things and do everything we can to make sure that our lifestyles are in harmony with the world.

Transformation Is Feasible

In October 2018, the Stockholm Resilience Centre released a study called "Transformation Is Feasible – How to achieve the Sustainable Development Goals within Planetary Boundaries" (Collste et al. 2018) which sought to understand how to best achieve the 17 United Nations Sustainable Development Goals (SDGs). The premise was that while many governments, corporations, individuals, and organizations like ours were interested in addressing the 17 goals, no one had thoroughly examined the underlying economic foundation to strategize how best to achieve them.

For the study, the Centre decided to model four different scenarios and try to see which ones would result in the most SDGs being achieved. The first model was called SAME, which proposed business as usual without any changes. The second was called FASTER and focused on trying to grow our economies as quickly as we can, ramping up GDP to have lots of money to be able to spend on SDG pursuits. The third model was called HARDER and called for heavy investment of resources toward achieving the SDGs as fast and as completely as possible. The fourth model they looked at was called SMARTER or Transformational; this model called for a radical shift in how we do things in the world, what we consider to be good and what we consider to be successful. The SMARTER model proposed completely different ways of working and new metrics for measuring success.

The Centre modeled the four different scenarios through the year 2050. Of the four, they found that the SMARTER scenario resulted in achieving the most Sustainable Development Goals. When they examined each scenario's effect on the planet, they found the SMARTER scenario not only stopped the trajectory of destruction to the planet but also actually started to reverse some of the damage. None the other three models SAME, FASTER, or HARDER could keep us from crossing the red line for planetary distress.

The report was clear: to ensure success in achieving the most Sustainable Development Goals and to steer the planet away from destruction, we need to develop and adopt bold transformational policies. Our current metrics of profit and loss, return on investment and even triple bottom line are so insufficient to rise to the present moment that they are rendered irrelevant. We must come up with a different way about thinking how we interact with the world and how we operate our organizations.

Regenerative thinking

Lifestyle change isn't easy. Nor is changing the way we think about the work we do. But there are promising ways to get started. A good roadmap for transformation can be found by embracing Regenerative Thinking. One of the best ways to understand regenerative thinking can be found in the work of author, educator, and regenerative practitioner Carol Sanford, who contextualizes regeneration by identifying four different paradigms of how we can interact with the world – paradigms which not coincidentally find significant parallels in the models examined in "Transformation Is Feasible" (Sanford 2019).

Sanford first presents the EXTRACTIVE paradigm. A person who exists in this way of thinking sees the world as fragments there for the taking. It's all about "me"; a person with an extractive mindset takes what they need without awareness or consideration for who or what is hurt in the process. Colonialism constitutes an extreme example of this paradigm, and versions of it remain dominant in most Western societies.

The next paradigm is called LESS BAD. A person with a LESS BAD mindset still sees the world as fragments but now is concerned with stabilizing them and reversing the damage. It's not just about "me" anymore; in a LESS BAD paradigm, there is some recognition of interconnectedness. LESS BAD is where the environmental movement started back in the 1970s, with the call to action of reduce, reuse, recycle.

The third paradigm is called DO GOOD. In DO GOOD, the world is still seen as fragmented, but there is motivation and understanding that the fragments can be improved; the idea of reciprocity comes to the fore. This paradigm sounds good at face value, but it falls short of embracing a shared understanding of our existence and requires a person to determine what good is and then impose that on other people, which fails to take into account the unique essence of their institutions, communities, and bioregions.

The last paradigm is called REGENERATIVE. With a REGENERATIVE perspective we do not see the world as fragments; we see it as wholes, and wholes within wholes. We see nested systems and perceive everything as being connected, with our decisions and strategies arising from an awareness and reverence for the fact that each of us is acting in relationship to one another, other species, and everything else on the planet. Regenerative Thinking evolves from a systems-thinking perspective to introduce that our role is not to merely recognize the system and stay out of its way but to actively add vitality, value and capacity – in short, to add life – to all other living systems. This transformational way of thinking is the way many indigenous people see the world. It is a perspective with a long history on this planet, and embracing it is a form of reawakening.

Sanford expands on her definition by introducing seven key principles of regenerative thinking, four of which are particularly important to museum/cultural work. The first one is the idea of ESSENCE, that every person, every species, every individual within a species has its own essence. Everyone and everything in this world is unique. The second principle is the idea of NESTEDNESS, which is the recognition that we exist within lesser and greater nested systems and that they are all interconnected with each other. The third one is RECIPROCITY, where we recognize that we operate within living dynamic systems, and we can find these relationships that are mutually beneficial. The fourth idea is DEVELOPMENT, where we recognize that everyone and everything is trying to evolve and develop their capacity and potential, sourced from their own individual essence. To engage in regenerative thinking is to recognize and support that development.

Phipps and the four paradigms

Phipps provides an excellent example of an organization evolving and moving through the four different paradigms of EXTRACTIVE, LESS BAD, DO GOOD, and REGENERATIVE. Constructed in 1893 as a gift to Pittsburgh by steel magnate Henry Phipps at the height of the last industrial revolution, Phipps Conservatory emerged as a manifestation of the idea that there was no limit to the amounts of resources we used or the pollution we produced, and that people could and ultimately would conquer nature. The Victorian style glasshouse, the largest in the United States at that time, was an oasis of green in a city engulfed in factory smoke and soot. Visitors flocked to Phipps to enjoy the flowers and fresh air, and so began a tradition where people would line up to enjoy the renewal of its seasonal flower shows and opportunity to deepen their connection to nature, which forms the bedrock of the organization's efforts to this very day. Today, Phipps remains one of the largest conservatories in the United States, and the original spring and fall seasonal flower shows have expanded to year-round, generating an annual attendance of around 500,000.

Following the institution's privatization in 1993, Phipps evolved from this EXTRACTIVE paradigm to become LESS BAD, with an emphasis on mitigating harm; this paradigm shift resulted in the design and construction in 2005 of the first LEED® certified visitor center in a public garden, LEED® Platinum production greenhouses, and one of the world's most energy-efficient conservatories. LEED certification, a then-new standard for green building design and construction which focused on building and energy efficiency, not only taught us much about what building greener could look like but also inspired us to go above and beyond what was required and then considered possible and to apply those standards in everything we do.

For its further expansion plans, Phipps initially pivoted to a DO GOOD paradigm, placing an emphasis on producing more energy than we consume and providing a healthy place for staff, before ultimately embracing REGENERATIVE thinking, in which we see ourselves in an interconnected relationship with everyone and everything on the planet. This change in perspective shaped the construction of three of the greenest buildings in the world on the Phipps campus. In 2012, Phipps constructed the Center for Sustainable Landscapes (CSL), the world's only building to achieve seven of the highest green construction certifications: the Living Building Challenge, LEED Platinum, WELL Building Platinum, Zero Energy Certification, SITES Platinum, Fitwel 3 Star, and BREEAM Outstanding In-Use. In 2015, the CSL was joined by the Nature Lab at Phipps, a Petal-certified net-positive energy modular classroom built with nontoxic materials. In 2019, the

Exhibit Staging Center, a net-positive energy adaptive reuse of an old public works building, was opened; it would go on to achieve Living Building Challenge, LEED Platinum, and WELL Platinum certifications. Phipps now hosts three Living Future Institute-certified buildings, each representing a unique construction type – new, modular, and adaptive reuse – and each demonstrating the potential of regenerative thinking and design.

These three buildings finally reach the REGENERATIVE summit by doing more than mitigating harm – together they foster an ecological rebirth of the former brownfield on which they are built. They are based on understanding the essence of Phipps and our community and developing strategies that recognize our relationship with the nested systems of the life shed in which we are embedded. By generating more energy than they consume, they not only achieve net zero but also add back to the grid. By hosting over 100 native plants in their landscape, they provide habitat and food sources to local wildlife. By incorporating a system of on-site rainwater processing, they ensure that rainfall on site stays on site rather than further taxing the region's combined storm and sewage system. By offering fresh air and views of nature to their occupants, they facilitate greater staff positivity and productivity. By opening their doors to hundreds of thousands of guests each year, they provide an inspiring example of building models for communities, businesses, schools, and homes that seek to make positive integrations with the natural world. In every sense, they operate in the direction of greater capacity and life.

In creating places that are good for people and the planet, the Phipps team discovered the need to go beyond buildings and lead by example by initiating a holistic rethinking of how we interact with the public and the world. Today, regenerative values influence every facet of Phipps, from buildings and operational practices to programming.

Living Systems Stakeholders

At Phipps, to help us understand how to use regenerative thinking, we use several different frameworks to guide our thinking. The Living Systems Stakeholders framework (Sanford 2011) for defining one's five principal stakeholder groups was originally designed for for-profit corporations but is equally effective for nonprofit organizations. The five stakeholder groups are defined as investors, users, community, co-creators, and the planet.

For a nonprofit organization, the investors group could be defined as your funders. For museums affiliated with a university, it might be university administration; for a government agency, it's the government that you are part of. The investors can be any one of these different entities. Most corporations/organizations principally serve only their investors. Some corporations might also serve their users. Fewer still might serve the community, and even fewer will serve their co-creators – which are their staff, their suppliers, and volunteers. Historically, the planet has been the most ignored stakeholder group.

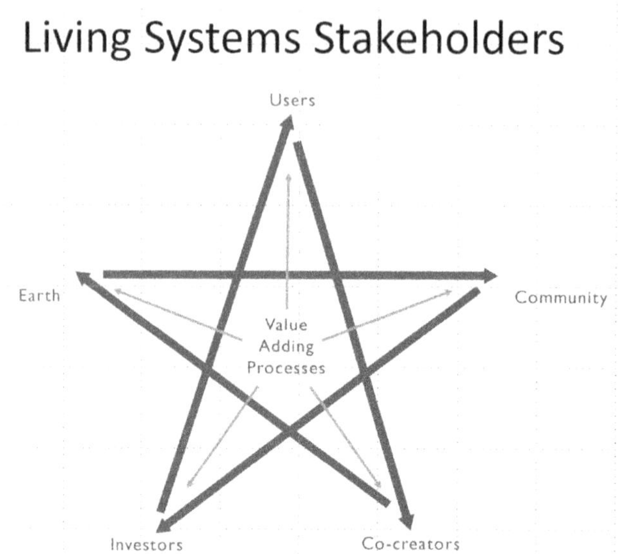

FIGURE 12.1 The Living Systems Stakeholders framework

In regenerative thinking, we seek to make sure that we are adding equal value to all five stakeholders in everything that we do. We are not just focused on investors or users, staff, or community. We want to make sure that everything we do is adding value to all five stakeholders, and when a project or program falls short of doing so, we want to explore how that program can be modified for greater capacity building across the stakeholder spectrum.

Three Lines of Work

The next framework we use is called Three Lines of Work (Regenesis Institute for Regenerative Practice n.d.). In this framework, we hold three different levels of engagement in mind at one time – starting with self and expanding outward to a greater whole (such as the organization) and then further outward to an even greater system we are trying to transform (such as the community).

In one example of the tool, you can consider an organization like Phipps, or another museum; in the center, we might hold each staff member. We try to make sure that staff understand what their job is. We help them to be proficient in what they were hired to do, and be the best that they can, but we also help them to see their relationship with the second line of work, which is the organization as a whole. We want them to get out of their silos to understand the mission, principles, and goals of the organization and how they play a vital role, along with every other

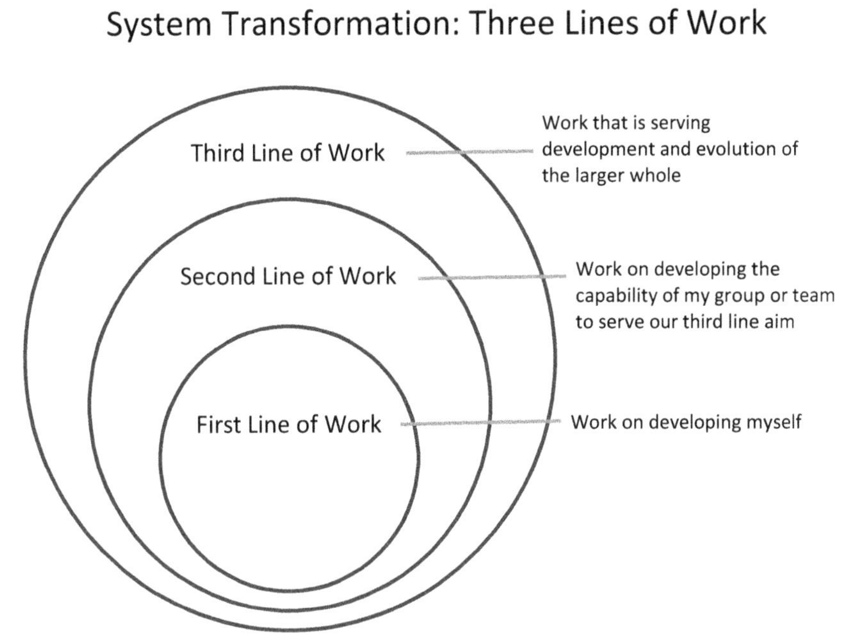

FIGURE 12.2 System transformation: Three Lines of Work

department, in meeting them. Next, we focus on our third line of work, which in this example is the community. What work could we be doing collectively to help our organization serve in the development and evolution of the community? What could each individual be doing to try to meet this aim? We encourage people to constantly consider the potential effect we can have in our community and to understand that individual actions can have collective, far-ranging impact.

Law of Three

The next framework that we use is called the Law of Three. The Law of Three is based on the idea that when an activating force (1) comes up against a restraining force (2), there must be a resolution (3) of either compromise or reconciliation. Because this situation is typically seen as conflict, the solution that most people try to use is to try to reach a consensus or a compromise. Consensus and compromise can often mean that everybody loses something. Instead, with the Law of Three, when an activating force comes up against a restraining force, we try to see if there is a way to reconcile it into making something better.

A good example of this at Phipps was the way we reacted when we first encountered the COVID-19 pandemic in March of 2020. At Phipps, we are primarily an indoor conservatory experience, and this put us in the situation that was more in

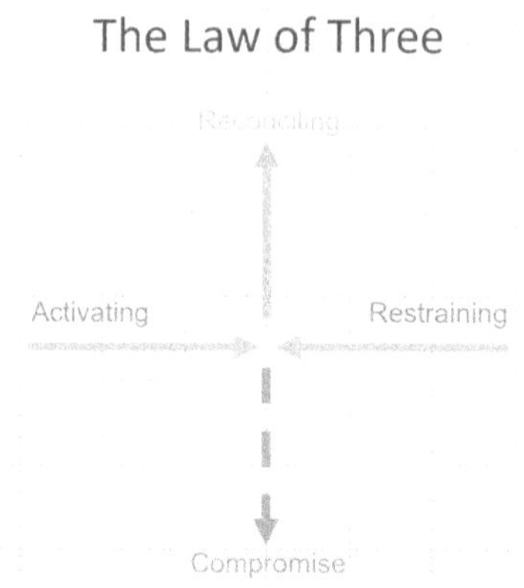

FIGURE 12.3 The Law of Three. IDP, Inc. 2002

common with indoor museum experiences than outdoor gardens. We had to restrict attendance to maintain social distancing. As a result, we saw our attendance not only drop by about 40% during the first two years. But also, during the first three months of the pandemic, we were forced to be closed. When this happened, a more traditional business reaction might be to layoff or furlough staff, which is what many of our peers did as they began to shut down. For Phipps, laying off staff would have been compromising in the face of the restraining force.

Instead, we looked for ways to take advantage of the restraining force and examine the impact this would have on our organization. In reconciling that conflict, we started by providing a safe environment for essential staff at the conservatory to keep all the plants alive and keep the facilities running. For other staff there were remote work options, and for the majority of staff who might otherwise have been let go, we looked for new areas of work that could put us ahead and make us stronger in the long run.

We had staff work on new exhibits and new programs and initiatives that we never had time to accomplish in years past. These included updating our standard operating procedure manuals, our databases, our plant collections, our photo archives, and our historic files. We had one staff member who worked on completing our American Alliance of Museums accreditation. We started taking all our classes and programs virtual. We had some staff call all of our 18,000 members. Others completed their required continuing education credits. And of course, we made thorough plans for how to have a safe reopening; when we were permitted to reopen three months later, we were able to do so on the very first day with all our staff on board and engaged.

Throughout the pandemic, we have continued to maintain our goal of adding equal value to all five of our stakeholders, including all our initiatives to support the planet. We found that we were in a stronger and better position after the pandemic hit than we were before because of all the extra work the staff accomplished that we never had the time to do before. The Law of Three gave us the means to see restraints not as problems, but as opportunities, and to use those opportunities to create something better than what we had before.

The long-term view

Another important context for regenerative thinking is that it demands a long-term view about the way we do things. Without this as a frame of reference, we are stealing the future from the next generations. Typically, organizations are overly focused on immediate return on investment, and this narrow mindset misses the bigger picture. Phipps is a nearly 130-year-old organization, which serves as a reminder that we need to make decisions that are not just focused on the immediate future, but rather put the organization in the best situation for success over the next 100-plus years. The smartest strategies we adopt are the ones that will make us successful over the long haul – and this is true of any institution with an interest in its own longevity, no matter how young.

Our short-term myopic thinking is especially visible in building projects. The Packard Foundation completed a study in 2002 (BNIM Inc n.d.) that compared the construction of a market-rate building to ones that were built to progressively higher green building standards starting with LEED Silver up to Living Building Challenge certification, which is the most rigorous green building standard in the world and includes Net-Zero Energy and Water requirements. Note that this study was done in 2002 when the difference in price between a market rate building and a green building was significant. The price differential has dropped dramatically since then as more and more companies and designers have become familiar with the concepts.

The True Cost of Conventional Building:

Operational and Maintenance Expenses

BUILDING TYPE	UPFRONT DESIGN + BUILD COST	NET PRESENT VALUE		
		30 YEARS	60 YEARS	100 YEARS
Living Building	$16.6 m	$18.7 m	$19.6 m	$20.8 m
Market Rate Building	$12.6 m	$22.7 m	$62.9 m	$348.9 m

FIGURE 12.4 The true cost of conventional building

In the study, if we were to compare a $12.6 million market-rate building to a same-sized building built to the Living Building Challenge standard ($16.6 million), we would see an initial construction cost difference of $4 million. Many owners, based on traditional metrics such as return on investment or payback period, would select the market rate building, consider themselves savvy investors, and keep the initial $4 million in savings. However, if we look at the cost to operate that building over the next 100 years, we see a significant advantage with the Living Building Challenge building; over 100 years, that market-rate building would cost $348 million to operate compared to $20 million for the Living Building Challenge building, owing most significantly to electricity, water, and natural gas costs coupled with the need for replacement of equipment and materials (as market-rate buildings are currently designed with a life span of 40 years compared to an expected 100 for Living Buildings). For a museum, that means over the next 100 years, there would be an additional $320 million to put towards programming rather than paying for energy and other expenses to run a building. In addition, this is before factoring in the amount of environmental damage that would be avoided and the amount of ecosystem services that would be provided by choosing the Living Building Challenge which adds even greater value to the decision.

We need to get beyond focusing on immediate payback and start thinking about what the impact is on our institutions, and on each and every one of our stakeholders, over the long haul. This kind of thinking requires an entirely new frame of reference – a new lens through which to see the world. I will not be around 100 years from now to see the impact of my decisions, but as the CEO of this institution, it should be my responsibility to make sure that my organization is in the best possible position to succeed in the future. That is the way we must be thinking.

Success through a new lens

When you change the lens by which you view success, you will see that you can accomplish many more things that you probably would have thought were impossible before. Since the time when we built our first LEED certified building, in 2005, Phipps has switched to using 100% renewable energy, built three zero-energy buildings, divested from fossil fuel investments, offset all the carbon heating our buildings, and stopped selling disposable bottled water, to name just a few dynamic changes.

For some of these initiatives, like purchasing renewable energy, there was a slight extra cost involved which we accept as a fair and reasonable cost for doing things the right way. For other initiatives, like not selling disposable bottled water, we are forgoing money that could be made – but here again, we accept that this is the cost of doing the right thing, and we work to figure out other ways to make up for it. In a long-term view, the money we would save or make by not doing things the right way will always reappear in another form later; the savings we would extract from purchasing conventional energy today become CO_2 emissions tomorrow, and the bottles of water we would sell become harmful waste.

Our Evolving Thinking and Action

2005	- First LEED Visitor Center in a Public Garden
	- 100% Renewable Electricity Campus-wide
2006	- No plastic disposable service ware
2009	- No disposable bottled water
2010	- Offset all carbon produced to heat all our buildings
2011	- No soda and no junk food in café
	- No factory farm meats
2012	- First Zero-Energy Building
2015	- Second Zero-Energy Building
	- Divested from fossil fuel investments
2017	- Switched to ESG Socially responsible investments
2018	- Third Zero-Energy Building
	- Focused on reducing single-use plastic
2019	- Greater than 70% of Café menu is vegetarian/vegan
2022–2023	- 40% reduction in fossil fuel steam heating

FIGURE 12.5 Our evolving thinking and action

What good looks like

What does good look like? For us at Phipps, the Center for Sustainable Landscapes (CSL) provides a strong answer to the question. The building turned ten years old in 2022, and at that time, it was still the only building in the world to have met seven of the highest green building certifications.

The CSL is a building that is good for people and the planet. Its building materials contain no toxic chemicals. It incorporates natural ventilation, natural daylight, and many opportunities to connect to nature. It was entirely built with off-the-shelf technology. It demonstrates that building the greenest building in the world is not limited by capability or technology, as long as we have the will to change the way we think about what good looks like.

Crucially, the CSL also highlights something the environmental movement got wrong so many years ago. When it first took off in the 1970s, the movement's emphasis was on making people focus on sacrifice, but this led everyone to think that if you wanted to be sustainable, you were going to have to give things up, and that, inevitably, you were going to be uncomfortable. Nobody aspires to be uncomfortable. If we want people to change, we must show them that the new way is better. The Center for Sustainable Landscapes is a much better space than almost any other that people might encounter for working, living, or learning. Museums, gardens, zoos, and other cultural institutions are uniquely trusted organizations, and it is incumbent upon us to show people that a better world is possible.

With all the evidence in favour of doing good, what prevents us? One of the biggest barriers is falling victim to the status quo and becoming set in our ways. We must see our way beyond that barrier, and this is where regenerative thinking can help us; in order to find the will to do the right thing, we must begin by changing the paradigm, the framework, and the lens with which we see the world.

Reflections on COP

Each year, heads of state, ministers, negotiators, climate activists, mayors, civil society representatives, and CEOs meet to attend UN Conference of the Parties to the United Nations Framework Convention on Climate Change (COP) to deliver action on an array of issues critical to tackling the climate emergency, from urgently reducing greenhouse gas emissions, building resilience and adapting to the inevitable impacts of climate change, to delivering on the commitments to finance climate action in developing countries.

I attended the COP26 meeting in Glasgow in November 2021. There was an impressive number of delegations from all over the world. What struck me, however, was that the overwhelming focus was on top-down solutions. There was little regard for the need for and importance of individuals to take and demand action. There were very few museums or other cultural institutions represented at the meeting. It is vital that we understand that governments are not going to solve the climate crisis alone; without grassroots support to pave the way, governments and their policies can be removed and replaced in the blink of an eye. One day your nation is pledged to support the Paris Agreement, the next it has pulled out.

It is going to take a combination of both top-down and bottom-up solutions to make the changes that we need to change the way we are living in this world, and museums can play a critical role in this effort. This is even more important when we consider how the courts and politicians influenced by lobbyists with unlimited resources are trying to impose laws against addressing climate change and weakening those that already exist. Under such a scenario, engaging the public to demand action becomes even more critical, and modeling solutions in our own institutions becomes even more important.

Urgent climate action is the focus of the UN Sustainable Development Goal 13, and climate change relates more or less explicitly to all 17 SDGs. Museums are specifically mentioned in the Workplan for the Paris Agreement (which was agreed to by the countries that are signatories to the Agreement in 2018), acknowledging the important role that museums can play in "enhancing the implementation of education, training, public awareness, public participation and public access to information so as to enhance actions under the Paris Agreement" (McGhie 2019). Yet, most museums are unaware of this.

There are over 103,000 museums (Statista Research Department 2022) around the world that reach hundreds of millions of visitors each year. Few other sectors can match the combined reach and level of trustworthiness that we have. Imagine if each of us were engaging our visitors by leading by example in helping them to become proficient in understanding climate change and the things that they can do to address it.

Now imagine the impact if we were to work together with other collections-based institutions such as gardens, zoos, and aquaria. Together, we represent some of the most trusted organizations in each of our communities. People trust us and come to us for inspiration and learning. They expect that we will be telling them the truth. Imagine if we were to use this opportunity to engage them in addressing climate change.

It is with this potential in mind that Phipps started The Climate Toolkit, which is working in partnership with the American Alliance of Museums, American Public Gardens Association, and Botanic Gardens Conservation International: to create an opportunity for museums, gardens, zoos, and aquaria to start working together and to share and learn from each other. Why should any institution have to reinvent the wheel? Let us work together and share what we are doing.

With the Toolkit, we started out by identifying nine different areas that were related to climate change: energy, water, waste, food service, transportation, investments, landscape and horticulture, research, and internal and external engagement. Then within those nine different areas, we identified specific actions that organizations could adopt to address climate change.

Initially when we developed the Toolkit, we were approaching it from a DO GOOD paradigm. We worked with other gardens to come up with a list of actions that we felt other gardens should do. The idea was that everyone should strive to do all these initiatives that were on the list. We soon realized that a DO GOOD model will not work, because what works for one institution may not be exactly right for another.

As we began to move it closer to a regenerative paradigm, we recognized that just as every individual is unique, every organization is unique, and every community and country is unique, and the Toolkit must support that understanding. Since that shift, we have seen more interest in institutions joining the Toolkit. We now have 119 institutions representing museums, including art, history, and natural history museums as well as gardens, zoos, and aquaria. Most of them are from the United States where the Toolkit originated, but increasingly we have institutions from many other countries joining as well.

The Toolkit allows organizations to identify action items that make sense for them and their community and then sort the Toolkit by that particular category to find other institutions that are working on that activity. They can then contact those organizations to ask such questions as "How did you do it? How did you convince your board? What do you recommend we do or not do?" If you are in a desert area, maybe water conservation issues are the most important to you; if you are in a more temperate climate, maybe reducing or using renewable energy is more important. For food service, how can you switch over to a mostly vegan and vegetarian menu? How do you get rid of bottled water and reduce waste? For transportation, how can you support better ways for staff and visitors to travel to your venue? For investments, how do you convince your board to divest from fossil fuels? How do you engage your staff? How do you engage your constituents? What types of research are being done? Every museum not only is unique but also has a different amount of resources available to them. Every institution should have an opportunity to contribute to this effort and to share what they are doing with other institutions with similar means. As the Toolkit grows, the intention is to allow organizations to sort by bioregion or budget size to find institutions facing similar conditions and restraints.

Because young people are the most motivated and critical actors in the effort to address climate change, we also created a Youth Climate Advocacy Committee to get young people involved and support them. After a pilot year, we have begun to reach out to other institutions that want to engage and support the young members in their communities. Ultimately, this effort will provide a forum through our institutions to engage young people from different cities around the world to address climate change. When you think about young people, they are the ones that are going to inherit the Earth we leave them. They have the most motivation to do things. They are our future members and our future donors. Let us get them involved now. Let us support them and show them that we are relevant and care about their future.

Conclusion

We need a completely new way to address climate change and understand the pivotal role that museums can play. This must start with a change in the way we view our place in the world and adopt a transformational rethinking of the ways we measure success and shift our focus to taking a long-term view on the impact of our actions. This should be followed by committing ourselves to adopt lifestyles that will allow everyone in the world to share in the Earth's limited resources and then by leading by example in our communities by encouraging our constituents to join us. This engagement will help create the bottom-up support that will be needed to make top-down policies and action possible. There is no other sector in the world that is as poised to lead in this effort as museums. By working together and sharing and learning from each other, we can unleash a powerful force of good to address the most critical issue of our time.

References

BNIM, Inc. n.d. *David and Lucile Packard Foundation Sustainability Matrix*. https://www.bnim.com/project/david-and-lucile-packard-foundation-sustainability-matrix

Collste, D., Cornell, S. and Rockström, J. 2018. *Transformation Is Feasible: How to Achieve the Sustainable Development Goals Within Planetary Boundaries*. Stockholm: Stockholm University.

Global Footprint Network. 2022. https://www.footprintnetwork.org/
McGhie, H.A. 2019. *Museums and the Sustainable Development Goals: A How-to Guide for Museums, Galleries, the Cultural Sector and Their Partners*. Liverpool, UK: Curating Tomorrow.
Regenesis Institute for Regenerative Practice. N.d. https://www.regenerat.es/
Sanford, C. 2011. *The Responsible Business*. Hoboken: Jossey-Bass.
Sanford, C. 2019. *A White Paper on Regeneration's Significance – Part 2: The Four Modern Paradigms*. https://carolsanford.medium.com/a-white-paper-on-regenerations-significance-part-2-the-four-modern-paradigms-ef306f622d1d
Statista Research Department. 2022. *Global Number of Museums 2021, by UNESCO Regional Classification*. https://www.statista.com/statistics/1201800/number-of-museums-worldwide-by-region/

13

CREATING ADVOCATES FOR THE PLANET

Clare Matterson

An interesting time to create a new strategy

Housing one of the world's greatest natural history collections, which documents and maps the evolution of species and change over time, in 2018 the UK Natural History Museum asked itself 'what is the role of a natural history museum in the 21st century?'

Dating back to the founding of the British Museum in 1753 and envisaged as a 'cathedral to nature' by its first Director, Richard Owen, the NHM today is home to 350 scientists and welcomes in the region of five million visitors to its South Kensington site and reaches 15 million people online.

As its previous strategy was ending, it was time to build a new future for the Museum at a moment in time when scientists were describing a new geological era – the Anthropocene. An era where humans were for the first time the single most influential species on the planet, causing significant changes to the land, oceans, atmosphere and organisms. Knowledge about the scale of disruption caused by human activity was growing thanks to the work of scientists across the globe, including those at the NHM. It was clear that this disruption to the planet's natural systems was so substantial that it was threatening the gains made over the past century to human well-being and health.

At the very moment the NHM began to consider its own future and role, there was growing public discourse and demonstration about the state of the planet. On 17 November 2018, in what was called 'Rebellion Day', about 6,000 people took part on a coordinated action to block the five main bridges over the river Thames, described by *The Guardian* newspaper as one of the 'biggest acts of peaceful civil disobedience in the UK in decades'. Greta Thunberg caught the public imagination when she started spending her Fridays outside Swedish Parliament to call for stronger action on climate change and then addressing the 2018 United Nations

DOI: 10.4324/9781003347606-15

Climate Change Conference. And quickly, young people protested collectively under the movement 'Fridays for Future'.

For a Museum that attracts large numbers of young people and is fundamentally about the natural world, these were events that could not be ignored. The UK Government had pledged to reach net zero by 2050 and set a target to cut carbon by 68% by the end of the decade – achieving such targets required the understanding and support of people to change their behaviour. It was clear to everyone at the NHM that addressing the climate and ecological crisis would require significant changes across society. The NHM was waking up to its role in helping people react to the scale of change. Over the next year, it would address the fundamental question 'How should a natural history Museum respond?'

What did audiences want from a natural history museum?

A key task was to understand how museums, and the NHM in particular, were already perceived. Information and data were collated from a range of sources and we commissioned independent research from the Museum and Gallery Insight Consortium on public (n = 1,500) and industry/corporate stakeholder's perceptions of the NHM (report not published).

The NHM already knew from its existing internal evaluations and visitor surveys that it was described consistently as being trustworthy and reliable but not considered to be innovative or interactive. Visitors reported in frequent evaluations that they saw NHM as a place about the past, and that they would like it to go further and be more relevant to everyday lives.

Public survey data all pointed to an increased recognition of and concern about climate change. When comparing the 2016 wave of the European Social Survey[1] to NHM's independent 2019 data, there was a marked increase in the proportion of people who attributed climate change to human activity. Similarly, in response to the question, 'How worried are you about climate change?', levels of concern had increased, with over three times as many people describing themselves as 'extremely worried' in 2019 compared to 2016.

When asked about the role that a museum should play in society, there was a strong focus on education and learning among all stakeholders. While they recognised the multifaceted work of museums within society, they were clear that education remained a priority.

> The NHM exists to educate and develop a deep understanding of all of its subject matter.
>
> **(Corporate funder)**

> The NHM exists to educate the public on natural history.
>
> **(Philanthropic funder)**
>
> Its primary function should be to educate, to do that it's a repository of information . . . the natural world in one hundred years or whatever that will be the only place that a lot of children will ever see anything. It's a particularly important place within the museum world because of the state of the natural world.
>
> **(Science & Engagement Funder)**
>
> Quotes from 2019 NHM Audience Perception Research

When we asked about the 'main purpose' of a museum, the most important function for all stakeholders was its collection – caring for our heritage, protecting and displaying their collections.

Yet, underlying these common themes were marked differences in attitude across generations. Although all considered that the purpose of museums was to act as guardians for history and heritage, this was much more marked for older adults (50+). Younger adults aged 35 and below were more open to museums playing an active role – changing minds, inspiring action and reflecting global issues.

In response to the question 'whose responsibility is it to care about the planet', the public allocated most responsibility to the government and schools but placed museums ahead of the media, public figures and charities indicating that museums can have a role in encouraging concern for the planet. However, only a small minority expected to see museums engaging in environmental debate, despite its prominence at the time in public discourse.

Shaping a new strategy

These findings provided the NHM with challenges and opportunities. It was apparent that stakeholders saw the work of the NHM in broad terms – education and preservation, rather than contributing to contemporary issues. However, there was appetite for change, particularly among younger audiences that museums should have a role in changing minds through their education and scientific work, although this might be unexpected for many. The opportunity was to explain and articulate the relevance of the collection and science on topical issues, rather than simply displaying the collection as a set of artefacts from the past.

The actual process of developing the strategy was collaborative, iterative and evidence-based from the start. Led by the NHM Executive team, it was a consultative process with involvement from trustees and colleagues, as well as the testing of draft versions of the strategy with external stakeholders from science, business, philanthropy and the media. An early decision was to set a timeframe for the strategy to 2031 to mark the 150th anniversary of the opening of the South Kensington site and to align (almost) with the United Nations Sustainable Development Goals. Overall, it took about 15 months of discussion, drafting, consultation and debate to develop the final document, which was published in January 2020.

The result was a bold new strategy 'A Planetary Emergency: Our Response' that set out a vision of a future where both people and planet thrive. As set out in the document, '*to achieve this we will harness the powerful combination of our three key assets: our collection, our scientific work and our reach to a worldwide audience on our mission* ***to create advocates for the planet***'.

Crafting a purpose for the NHM

Both the vision and the mission require some explanation in terms of their development. When considering 'vision', there was consensus that the scale of the problem for humankind and the planet required collective action from people, governments and organisations. The fact that the NHM vision was in alignment with those of other organisations, such as the World Wildlife Fund (WWF), was seen as a strength as it is going to take the best minds on the planet and a global effort to tackle the planetary emergency. The view was taken that it was important for many major organisations to be dedicated to a common aim of a future in which nature and humanity co-exist sustainably, and it is only through mass collaboration and working towards a common goal that impactful change can be made.

The mission 'to create advocates for the planet' emerged from several different lines of thought.

First, the NHM is a collection of parts with people working in many different roles – from caring for the collection, running commercial retail, catering and event businesses, communicating science, managing the finances, doing cutting-edge science and others. It was important that all colleagues could feel connected to the purpose of the museum as everyone had a part to play in helping to secure a more positive future for humankind. Inspiration was drawn from the story of President John F. Kennedy visiting NASA for the first time in 1962 and casually asking the janitor what he did for NASA. The janitor replied, 'I'm helping put a man on the moon'. The janitor could connect his purpose with that of the organisation, and at the NHM there was a strong desire to create the same ethos for all colleagues with the new strategy.

Second, there was much discussion about the traditional role of the museum as the neutral, objective body set apart from the social and political worlds which they inhabit or, on the other hand, as an organisation with a point of view. The weight of scientific evidence about the state of the planet left no doubt that the NHM was ready to have a point of view. In fact, much of the science from NHM scientists was adding to the public record about the impact of human activity on the environment. For example the project PREDICTS (Projecting Responses of Ecological Diversity In Changing Terrestrial Systems), a database of more than 3.2 million records, sampled from over 26,000 locations and representing more than 47,000 species, is the largest of its kind and involves hundreds of scientists from around the world sharing research data on biodiversity. Dr Alex Bond, Senior Curator in Charge of Birds, had reported on unprecedented plastic ingestion in seabirds. Thanks to scientists at the Museum, all known plant species growing in the Amazon have been listed, underpinning all future studies on the rainforest's ecology and evolution. Given the role the Amazon rainforest plays in reducing the effects of climate change by absorbing vast amounts of carbon, this knowledge is vital to study and protect it for future generations.

Such was the scientific work of the NHM and the conviction about the perilous state of the planet drawn from scientific evidence, the NHM was confident to declare a 'planetary emergency' to set the context for the strategy:

> We face a planetary emergency. Humanity's future depends on the natural world, but we are not taking effective action to combat our destructive impact on the planet's survival systems. . . . We must act now, we must act on scientific evidence and we must act together. We need an unprecedented response.

Strong words indeed, but the question that remained was how the NHM should position itself in the context of the growing environmental activism. It was clear that it could no longer see itself as a passive bystander describing the state of the world. Yet, institutionally it did not see its role as an 'activist' as carried out by national and global NGOs. Rather, the NHM saw its role as being 'active' in focusing its research on seeking solutions for nature, explaining complexity and the interconnectivity of the planet and, crucially, on providing opportunity for learning and debate on the choices facing society.

Following this theme, the term 'advocate' resonated strongly through the NHM's ability to provide scientific evidence that would inform and help to shape choices. The big shift in emphasis was to use the collection to shape a positive future rather than it being a shrine to the past. The opening paragraphs from the strategy capture the sentiment:

> *Over the last century, humanity has benefited from astonishing improvements in health and in quality of life. Yet at the same time, we have been exploiting the Earth's resources in unprecedented and unsustainable ways. Our world is now changing fast under the influence of human behaviour: climate change and biodiversity loss are just some of the manifestations. These changes threaten the security of critical ecosystems and jeopardise the gains made over the past century in human health and well-being. The current trajectory will bring still greater threats. If we are to contain these threats and to enjoy, sustain and build on the advances of the past century, we need the knowledge and understanding that emerges from scientific discovery.*
>
> *We cannot hope to develop solutions to address problems such as biodiversity loss, food security and climate change without research that unlocks the underlying biology and applied science.*

Understanding life on our planet is the greatest scientific challenge of our age. But while this knowledge is necessary, it isn't sufficient. We also need to ensure it reaches decision-makers and informs and engages public audiences, so that they demand and support change, and adapt their own behaviour. Moreover, we need to offer people hope: hope that with the aid of good science and good policies, there is ***a future where both people and the planet can thrive****. Building and spreading this hope through evidence and action will be the foundation of our strategy in the coming decade. In today's challenging times, we will* ***create advocates for the planet****: citizens who feel sufficiently informed, confident and motivated to make wise decisions, to get involved, and to use their influence to make a positive difference to the global future.*

Activating the strategy during COVID-19

The eagle-eyed reader will have spotted above that the Strategy was launched in January 2020. Less than two months later, the NHM closed its doors to the public for the first time since World War II as a result of the COVID-19 pandemic.

So, how did the NHM activate its bold new mission to create advocates for the planet with its door closed and a vast number of its staff on furlough? The remainder of this chapter describes three areas of activity that took place during the development of the strategy and gained significant momentum during lockdown.

- Walking the talk – putting in place actions to measure and manage the sustainability of NHM operations
- Defining an 'advocate for the planet' – being clear on what was meant by the term and how to measure success and
- Activating the public programme so that public-facing activities spoke to the aims of the strategy

Walking the talk on sustainability

When publicly declaring a 'planetary emergency', the first place for external scrutiny must be the sustainability credentials of the declarer. The Strategy itself set out a clear commitment to sustainability,

We are putting environmental sustainability at the heart of our operations. We are committing the Museum to become the first museum in the world to set a science-based carbon reduction target in line with the Paris climate agreement 1.5°C global warming trajectory. A strong signal will be transmitted throughout the organisation that sustainability counts across all areas. We will make reducing carbon emissions in both our operations and our supply chain the central focus of our sustainability efforts.

The NHM was not only ready for the scrutiny but also painfully aware of the challenge ahead of it with an aging and dilapidated estate, caring for one of the most important natural history collections in the world requiring specific environmental conditions, along with the need to raise funds.

It started with the development of a 'Sustainability Action Plan and Commitments'. Created by an internal Sustainability Working Group that reported to the Executive Board, this plan reviewed the current state of sustainability efforts across all areas, highlighted the gaps and developed an action plan to keep the pressure up, set new targets and provide a clear platform for communication of work on sustainability.

The NHM set for itself the target to reach net zero by 2035 with science-based targets established to help drive down emissions as quickly as possible. It was recognised that some parts of the operation would always generate a small amount of carbon dioxide, no matter how energy efficient it became. So, once everything has been done to tackle emissions, to achieve true net zero, the NHM indicated that it would explore carbon offsetting schemes as a last resort.

Beyond carbon reduction, the plan set out actions across all areas of the NHM's work as described on its website:

- *Energy – We'll always need energy, so we want to source it sustainably.*
- *Supply chains – Many of our goals rely on cleaning up the Museum's supply chains. We want all our suppliers to join us in working in a way that's kind to nature.*
- *Waste – Carbon emissions aren't everything: pollution and overconsumption are two big threats to nature.*
- *Travel – Museum staff only travel when it's absolutely necessary because we know that planes and cars produce greenhouse gasses.*
- *Water – Water is one of Earth's most precious resources.*
- *Investments and income – Money is powerful, so we want to make our money work for good.*
- *Food and catering – The food we eat can have a big impact on nature. We want to ensure everyone who visits the Museum has a range of nutritious, sustainable food options to choose from.*
- *New buildings and exhibitions – All our new projects will have sustainability at their heart.*

The 'Sustainability Action Plan and Commitments' was published on the website under the heading 'Sustainable by Nature'. With a commitment to being open about its own sustainability journey, the NHM used clear and straightforward language to show the short- and long-term commitments and each year publishes a sustainability report to show the progress made.

Defining an 'advocate for the planet'

Expressing the ambition to 'create advocates for the planet' required a definition and action plan so that the term was not simply an empty cliché. In the Strategy, 'an advocate for the planet' is defined as someone 'who speaks up on behalf of nature, and ultimately will take action to protect it'.

Using insights from psychology, an internal NHM team explored the different elements of human interaction that are needed for people to gain the confidence to speak up for nature. Simply telling people that there is a problem is not effective. Instead, a first crucial step is a connection with nature. Such a connection is not developed simply through contact with the natural world. Rather, it is made up of emotional, cognitive and behavioural aspects – feelings about nature, knowledge and actions.

These insights were used to create the Advocate Engagement Model shown in Figure 13.1. The base of the model shows that underpinning any experience, some basic needs must be met as set out in the NHM Visitor Charter. Next, to inspire a connection with nature, it is necessary to create memorable and emotional experiences. Then, to inform about the natural world, relevance is important so that people place themselves in the story of our planet. Finally, to act for the natural world, people need to be empowered so they can make informed decisions about their own behaviour and actions.

The NHM team also developed three outcome measures to track change and impact:

- people act on behalf of nature
- people understand their role in nature and their impact on it and
- people feel connected to nature

This model was shared and discussed with all teams developing public programmes across NHM. The requirement was that whenever a new programme was being planned, it was critical to demonstrate how it met the aims of the model. Although

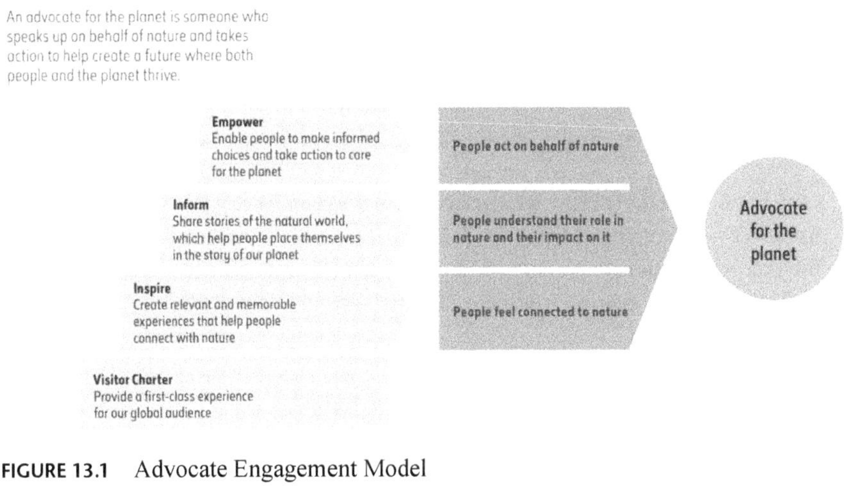

FIGURE 13.1 Advocate Engagement Model

not every activity would lead to action/empowerment, teams needed to be clear about the purpose of their activity in the context of the model. Each programme was evaluated using the shared outcome measures.

Activating the public programme to the strategy

As the Strategy was launched, an *Anthropocene Hub* was opened on the Museum's website packed with information, stories and actions about human's impact on the planet, including research on rain forests, corals and deep oceans by NHM scientists as well as actions that the public could take to protect nature. Aimed at the public, this evergreen web content has continued to grow with new stories and actions dedicated to sustainability and climate change.

Unknown at the time of the launch was that the NHM was about to enter one of its longest periods of closure in its history. In March 2019, with the new Strategy in place, the Museum was in lockdown providing a chance to stand back and think differently about what could be done to activate the Strategy. For anyone who works in a museum, it is common knowledge that to create a new gallery or an exhibition can take many years. Yet, the sense of urgency for more immediate action was palpable among colleagues and reinforced by the knowledge that COVID-19 was a product of the damaged relationship between the natural world and humans.

As a result, the NHM took a forensic look at its existing programmes and how they could play a role in creating advocates for the planet, as well as developing a new initiative. These are described as follows:

- **Urban Nature Project** – this existing programme was nationwide linked to the redevelopment of the NHM's five acres of outdoor space at its South Kensington site. With its focus on understanding change over time, biodiversity loss and with a very practical focus on developing scientific skills to measure and understand change, it was already well aligned with the NHM Strategy
- **Wildlife Photography of the Year** – this long-running global competition had already been shifting as wildlife photographers were witnessing the impact of humans on the planet. The NHM spotted an opportunity to use these images to connect emotionally to the beauty and fragility of the world to connect to new global audiences
- **Our Broken Planet: How we got here and ways to fix it** – this was a brand new initiative conceived and developed during lockdown to respond with urgency to the planetary emergency. Aimed at young adults, it sought to engage unflinchingly the state of the environment and the solutions developed by science and technology.

Each programme is described in more detail as follows.

Urban Nature Project – The UNP is using the gardens in South Kensington as a launchpad to work with partners across the UK to tell the story of life on Earth, raise awareness about human impact on the planet and empower people to protect urban nature. The ambitions of the UNP are to transform the Museum's five-acre gardens into an exemplar of urban wildlife research, conservation and awareness; to start an urban nature movement across the UK and to develop new scientific tools and skills that are urgently needed to protect urban nature.

The transformation of the South Kensington outside space uses stories of the past in unusual and exciting ways. When completed, as visitors enter the East Gardens of the NHM, they will be taken on a journey into the deep past and invited to explore the diversity of life as it evolves. They will learn about the explosion of life in the seas 500 million years ago, see dinosaurs grazing among tree ferns and cycads and track the evolution of our earliest ancestors. Visitors will become immersed in a landscape that gradually fills with plants, trees, reptiles, birds and mammals (including humans). By understanding changes that have occurred on our planet and how life responded, humans should be able to place themselves in Earth's story.

On the other side of main entrance, an urban environment of current times is encountered. The west lawn and Wildlife Garden open up to a living research laboratory, an urban space teeming with life, where the public can learn how to gather information about the changing environment. It will become a space to explore, reflect and connect with nature and find out about the extraordinary wildlife right on our doorstep. NHM scientists are actively monitoring and studying the biodiversity of the entire site, the results of which will be shared with the public.

Led by NHM scientists, urban nature professionals, academic researchers and conservation practitioners have come together to pilot a new range of technologies for monitoring change in urban environments. New techniques such as eDNA and acoustic monitoring are being tested and developed so they can be used by schools and the wider public. When completed, it will be possible to promote the NHM gardens as an exemplar of urban nature conservation and test-bed for science, share findings, learn from others, develop a partnership network and ensure that the science has real-world applications for urban nature conservation.

Finally, the transformation of the gardens is being developed as an exemplar in environmental sustainability with an aim to achieve net zero carbon, create a zero-waste garden and ensure no waste from the construction to landfill. Once open, the aim is to operate the cafés to high sustainability standards, grow the plants coming into the garden in the UK and highlight actions as part of the project.

In essence, the Urban Nature Project tells a story of change, shows science in action and illustrates simple essential steps the Museum as well as visitors themselves can take to become advocates for the planet.

Wildlife Photographer of the Year – WPY is the NHM's annual competition bringing the glories of the natural world to a global audience. What started out in

1965 as a magazine competition with 361 entries has become one of the world's most prestigious competitions with 50,000 entries and a touring exhibition seen by millions across the world.

While WPY has always toured to different museums, the Strategy gave the opportunity to reset the approach and narrative as well as extend to new venues. The photographers who enter their work into the competition have themselves become activists and conservationists as they have witnessed the degradation of the natural world that they are documenting. Therefore, it was decided to use this natural world reportage and its incredible soft power and emotional connection to turn up the noise and convene global conversations. By taking these powerful images, combined with quotes from photographers sharing their story and location, the NHM moved the WPY images from the gallery floor to global meetings, such as the World Economic Forum in Davos, to ignite conversations with global decision-makers on the issues facing nature today.

Our Broken Planet: How we got here and ways to fix it – The third and final example of bringing the strategy to life is perhaps the most innovative in how it is using the collection and science to engage with the public. With its new Strategy, the NHM saw itself as a critical player on the journey to global climate consciousness holding the key to wider public knowledge and debate, and it was time to kickstart a public debate about how and why our relationship with the natural world needs to change. Starting with the younger generation who were looking up to museums to bring about social change, the NHM wanted to create an experience to inspire them to make active, positive choices to improve and be critical of humankind's relationship with the planet. The plan was to create a programme of activity that directly addressed the very issues that young people had been protesting about as the NHM strategy was developed.

Conceived during lockdown, Our Broken Planet started with the NHM's collections, scientists and science. While everyone was working from home, NHM scientists were tasked to nominate an item from the collection, explain why this item told a story about the state of the planet and explain their research in seeking solutions to the problem.

What came back was an exciting array of stories, objects and science. Reduced down to 40 that best exemplified both the consequences of human actions and potential solutions, an exhibition and online conversations were born.

Human activity was allocated across three main themes: the food we eat, the products we use and the energy we consume. What transpired was an exhibition with these three themes opening in three phases during 2021, each representing a different system that contributes to the planetary crisis. In each themed section, a series of stories was revealed using the selected object and how the object told a story about today's environment, a picture of the scientist (and sometimes their voice) and their current research to seek solutions. The simplicity of the display, alongside the introduction to the scientist, created a personal and emotional response as part of the overall experience.

In advance of the physical exhibition opening, a series of online conversations between scientists, farmers, entrepreneurs, fashion designers, journalists, photographers and activists was launched to discuss the key challenges faced in building a greener future following the same themes. These are all available on YouTube.

In the exhibition, the 'Eating the Earth' section showed the impact of food choices on the planet. Alongside the stories about pollinators, fishing and agriculture, bright infographics compared the impact that different food have on the environment. For example that it takes 326.2 square metres to produce 1 kg of beef, and only 3.5 square metres to produce the same quantity of tofu. The next section, 'Nature for Sale', explored how the resources that make up everyday items including fashion choices and the use of smart phones can affect the planet. The final and last section, 'Climate Emergency', investigated how energy is used to power our world – and that even the renewable and alternative energy sources have an impact on our environment, requiring informed choices about the future use of the Earth's resources.

You won't hear any of these bees buzzing past you on a summer's day. All are extinct in the UK thanks to expanding farmland and the destruction of flower-rich habitats. Bee species around the world are declining, with serious consequences for us all. From almonds to tomatoes, 35 per cent of the crops we eat need animals to pollinate them. Bees are worth billions of pounds to the global economy.

Sadly, many of the UK's incredible 275 bee species are in trouble, but we can all help by planting wildflowers in gardens and window boxes.

Minerals such as lithium are vital for building batteries that power electric cars and store energy from wind turbines and solar panels. But there's a problem: switching every car in the UK to electric power would take three-quarters of the world's annual lithium supply. Faced with this challenge, Museum scientists are working out how to extract lithium from British granite mines – a source that could be much more sustainable than lithium from other countries.

(Examples of stories from Our Broken Planet)

Since opening in May 2021, Our Broken Planet welcomed over 750,000 visitors, including key climate activists such as Greta Thunberg;[2] its run was extended, and it has since been decided that it will become an ongoing major programme for the NHM for the foreseeable future.

Personal reflections

It is too early to know the impact and learning from these programmes and the wider Strategy. However, it is clear that the creation of the Strategy galvanised every corner of the NHM. The NHM had been doing cutting-edge science for decades from measuring the impact of climate change on species to helping discover new sources of metals needed for renewable energy. This Strategy marked an important shift in emphasis, from a Museum that reported on the world to one that actively participates in shaping solutions to global challenges. It includes a renewed focus on impact, through public engagement and science, in raising awareness of, and finding solutions to, the biggest challenges facing humanity and the planet. It sets the NHM as more than an observer and commentator on the natural world, to be proactive and inspire action.

Given the significant threats to nature from climate change and evidence that we are entering a sixth mass extinction – all caused by human actions, it will be exciting to see how the NHM continues to tread the line between activist and observer to use its incredible collection and science to achieve its new mission.

Notes

1 ESS Round 8: European Social Survey Round 8 Data (2016). Data file edition 2.2. Sikt – Norwegian Agency for Shared Services in Education and Research, Norway – Data Archive and distributor of ESS data for ESS ERIC. doi:10.21338/NSD-ESS8-2016.

2 https://www.nhm.ac.uk/press-office/press-releases/climate-activist-greta-thunberg-visits-our-broken-planet–how-we.html

14

THE WHITWORTH – TRANSFORMING MANCHESTER'S GALLERY IN THE PARK

Jo Beggs and Dean Whiteside

A history of re-invention

The Whitworth's story is one of continual reinvention. While change was often undertaken in difficult times, it has always been a dynamic organisation. Founded in 1889 as a philanthropic legacy of industrialist Joseph Whitworth, the Gallery was envisaged with a 'build it and they will come' philosophy. Beaumont, the original architect, undertook five projects over an 18-year period to transform a private residence, Grove House, into a public art gallery. In 1908, in a final flourish, he created the Gallery's grand and distinctive terracotta frontage and sweeping entrance from Oxford Road. Only during its creation did its contents begin to be assembled as the ambition of the founders inspired others in the city to help kickstart the Whitworth collection we know today.[1]

In 1958, the Whitworth became part of the University of Manchester. This gave extra weight to the educational mission of the gallery and prompted a pace-change in contemporary fine art collecting and a significantly raised status for the design collections, notably the textiles and wallpapers. As a very visible marker of the change, the University undertook a major remodelling of the interior, appointing John Bickerdike and Partners with a remit to create spaces fit to house the growing, cutting-edge, contemporary collection and to inspire and reflect the creativity and creative industry in the city. Drawing on European influences – in particular Scandinavian Modern – Bickerdike delivered a sleek open plan design with increased natural light, vast open spaces and two new grand staircases. In 1995, an architectural intervention was undertaken by ABK (Ahrends, Burton and Koralek) which created a large new internal gallery space, reclaiming an external light well at the centre of the building.

DOI: 10.4324/9781003347606-16

Strategic context for change

The Gallery's position on the edge of the campus brings it as close to its local communities of Moss Side and Rusholme as to the University. This provides a unique opportunity to be both an academic and public space, with its collections providing inspiration and sparking curiosity for a wide range of audiences. While rooted in the idea of a research institution, that research may be undertaken by academics or by five-year-olds.

The 2015 extension project had a number of driving factors. Primarily, it was to address growing visitor numbers, built over the previous few years – from 86,000 in 2006 to 139,000 in 2008 – and to provide space for an expanded range of activities, serving a more diverse audience.

There was a need to better protect the Gallery's collection of over 55,000 objects, comprising fine art, textiles and wallpapers, all stored on site. Storage had evolved over time as the collection grew, without any masterplan, resulting in 11 stores, all with varying environmental and housing conditions; from rolled wallpapers in award-winning bespoke storage on the first floor to textiles in cramped cupboards in the basement.

The Whitworth had always provided good access to collections, technically for all but in practice mostly for academic research. Credit should be given to staff for this previous level of access. Due to inappropriate conditions and poor physical access to stores, curators and members of the Collection Care team were regularly seen carrying large objects from stored collections up and down long flights of stairs to spaces where visitors could access them, the only lift in the building being a small passenger lift which connected the ground floor, mezzanine galleries and (for staff) the basement.

The 1960s' development had radically altered the three large spaces at the rear of the gallery to create spaces thought, at the time, to be more fitting for contemporary art. The Edwardian barrel ceilings had been hidden with a lowered flat ceiling, the void above providing the ideal space for a vast air conditioning system. This not only radically changed the volume of the spaces but blocked the natural light as well. One of the three spaces had been converted into a lecture theatre with tiered seating which, although ideal for lectures, had limited use.

The project also sought to re-introduce the building to its surroundings. Beaumont – in true Edwardian style – had paid no attention to the 'back' of the building, and Bickerdike had clearly seen no need to address it. At some point (probably in the 1980s when the park developed a reputation for being a dangerous place), an eight-feet-high fence was erected separating the building from the public park. This created a particularly urgent problem as the gallery sought to encourage increased use from its local communities. Visitors approaching from the park were met with the unwelcoming back wall (during the public consultation, one visit described it as 'like an old railway warehouse') of a building that did nothing to announce itself as a public gallery. Local people reported having lived close to the Whitworth all their lives but never having felt they were welcome inside. The project sought to change this, and a RIBA architectural competition was held in 2009 to select an architect whose values aligned with this.

A shared commitment to sustainability

'The brief for the Whitworth offered a unique opportunity to capture the qualities of the surrounding landscape' said Stuart McKnight, founding partner of MUMA (McInnes Usher McKnight Architects) who were selected to undertake the work. MUMA were, at the time, a relatively young practice with a limited number of completed projects, but for those they had delivered, such as a remodelling at Newlyn Art Gallery and plans for the re-imagining of the Medieval and Renaissance Galleries at the V&A (completed in 2009), they demonstrated a sensitivity and creativity that the Whitworth project needed – a commitment to sustainability and a holistic approach to the project that started with the needs of the organisation.

MUMA developed a scheme that involved an H-shaped extension to the unloved 'back' of the building, wrapped around the three large exhibition galleries and what would become new basement level stores. On the north/city side of the building, they created a new brick wing housing a large gallery space on the first floor with collection study room below, while the wing to the south/park is a light-filled masterpiece of glass and stainless steel that blurs the lines between interior and exterior. It houses the café with a learning studio below. Floor to ceiling windows along the promenade open up views into a new garden and across the park and flood the space with natural light.

It is fair to say that environmental sustainability, while considered a priority, was not as high a priority in 2009, when the brief was prepared and RIBA competition launched, as it would be if the project had been imagined a decade later, clearly highlighting the pace at which thinking in the sector around this issue has increased. Even so, the Whitworth set a high target for carbon reduction in the architects' brief, seeking to reduce carbon outputs by 10% despite increasing the footprint of the building by a third. Manchester is mindful of its industrial history and as such has ambitious targets to become a carbon-neutral city by 2038 (*Manchester Climate Change Action Plan 2020–25*), 12 years ahead of the UK Government target for the UK. The University of Manchester had also committed to delivering a greener, low-carbon campus in its strategic plan, and its Estates Team were particularly open to exploring new technologies that would support this. In addition, funding from the Heritage Lottery Fund (now the National Lottery Heritage Fund) came with expectations that all possible carbon reduction measures were at least considered. This included solar panels, wind turbines and grey water collection, but project budget and realistic building management projections meant that it was agreed to seek less infrastructurally intensive solutions and concentrate on how the specifics of the building and the site could be used to develop more passive solutions.

The project rationalised the 11 collections storage spaces into a centralised collections storage/care zone in the basement with level access to a new conservation studio, technicians' workshop, transit and quarantine rooms, a loading bay and goods lift. Despite a smaller footprint, bespoke efficient storage systems provided increased space for relocated collections with room for expansion. The wrap-around effect of the new building, environmentally buffered storage units and minimal air exchange rates help to keep the internal environment stable.

> *Our strategy was to work with the inherent characteristics of the building. By relocating the collection, within the inherently stable environment of the basement, with passive means we have created a stable environment for the collection. With no external walls (being wrapped by the new promenade) and benefitting from the substantial thermal mass of the historic fabric, temperature is stable.*
>
> *Humidity fluctuations are tempered by the use of lime plaster, with breathable paint to all walls and ceilings. Employing such means, together with a conservation heating strategy, has meant that we were able to remove air conditioning.*
>
> *(statement from MUMA architects in internal funding report 2015)*

The removal of the air conditioning system and the adoption of a conservation heating approach, without any chillers or de/humidification, made the largest contribution to achieving energy reduction targets. The existing system had become expensive to run, noisy and unreliable, causing fluctuations in temperature and humidity levels which were uncomfortable for staff and visitors and damaging for artworks. The Edwardian passive air system in the three largest galleries was re-established through the removal of the suspended slatted ceilings. Insulation was added, and the original ceiling windows were replaced with well-fitting, solar-shaded new roof lights. This action required extensive design discovery. It was selected on the basis of a forecast that it would deliver 16–28 degrees C and 30–70% RH (with no more than a 10% rate of RH change in 24 hours).

While major solar solutions were ruled out due to cost and the long payback period it would have required, a solar thermal array offsets domestic hot water demands in the kitchens and toilets while water restrictions are delivered through dual-flush toilets and aerators on taps. Consideration of grey water for toilets was ruled out due to the potential staining of toilet bowls which felt inappropriate within the clean design.

The new garden offered an opportunity to lay earth tubes beneath the lawn. These were to have the effect of bringing tempered outside air into the buildings to supplement the heating and ventilation systems. Despite being relatively low-tech, the earth tubes failed to deliver the expected results with minor faults having a significant impact on their effectiveness. Once the tubes were covered and the lawn laid, it proved impossible to repair and maintain the tubes without invasive investigation and disruption to public access to the Gallery. Sadly, they are a casualty of the timescale and budget restrictions of the project as a decision had to be made to accept the failure rather than attempt a disruptive and costly fix.

The reconfiguring of spaces not only helped to make sense of the interior layout, offering views to the outside, it also made best use of natural light, flooding the non-collection spaces and making good use of non-direct light sources in Gallery spaces and on the grand Edwardian stairwells. On the promenade, the floor to ceiling glass offers a bright space even on the dullest of days, and beautifully designed structural solar shading is employed to reduce glare (Figure 14.1). Energy-saving LED lighting was installed throughout after much testing for true colour rendition. Light levels were reduced where possible, and occupancy sensors ensure lights in back of house areas do not remain on when rooms are not in use.

The project made us aware of the quality of the materials used in the 1960s' development – in particular wood, slate and marble – that remained in good shape 50 years after installation. MUMA were committed to the use of materials of similar quality, which would not only deliver aesthetically but also be equally long-lasting. New materials were sourced as locally as possible (e.g. bespoke 'Whitworth Mix' terrazzo flooring developed in Leeds), while York Stone and cobbles from a disused service road in the basement of the building were reclaimed to create paths and walls in the garden.

In addition to structural and technological interventions, the project marked a change in the philosophy of the organization regarding carbon reduction. The Whitworth had long been looking to reduce emissions through robust recycling and reduction of food waste, promotion of low-carbon staff and visitor travel and

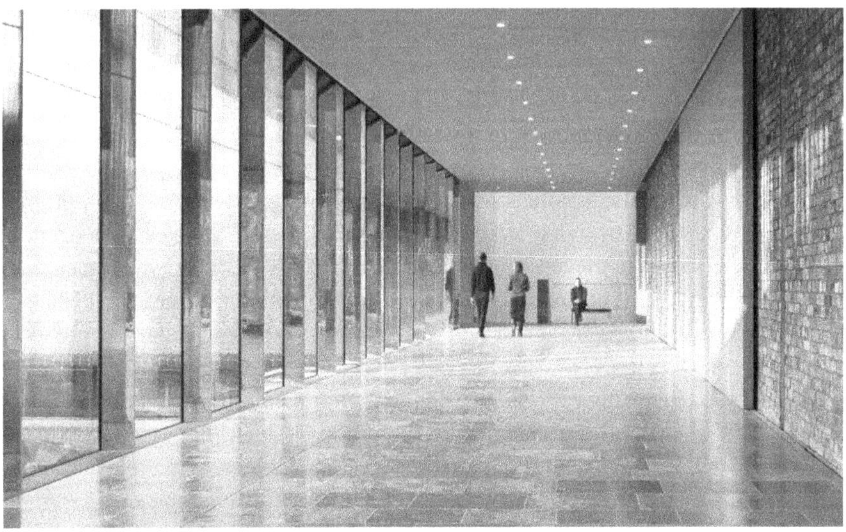

FIGURE 14.1 Promenade with exterior solar shading

attention to water usage. A Green Group had been established some years prior to the start of the capital project, and working with Julie's Bicycle, which supports cultural organisations in tackling the climate crisis, baselines for waste reduction, energy usage and water saving had been established and regular green audits undertaken. The capital project offered an opportunity to think afresh about carbon emissions as part of a much larger shift in organizational thinking. A significant amount of work was undertaken by the Collection Care Team, along with colleagues across the museum sector, to accurately understand the environmental needs of the collection, expanding the parameters for temperature and humidity control for more robust objects, both in storage and on display, and exploring the use of local climate control inside cases and frames for more fragile items.

Creating the new relationship between the Gallery and its surrounding green space offered a range of opportunities to promote biodiversity, engagement with nature and the benefits of the outdoors for good mental and physical health. Working with botanists from Manchester Museum, Gallery staff and local people helped undertake 'bio-blitz' audits of the park, with encouraging results. During closure, a member of the Learning Team worked in residence with a local Forest School, developing a programme that would lead to the creation of a new role within the Learning Team of Cultural Park Keeper, with a remit to deliver outdoor, nature-based creative activities for all ages, engaging local communities with their green space. Staff members were trained in bee-keeping, and a hive was established on the roof of the new extension.

Two gardens were created as part of the capital project: first, an Art Garden was situated between the two new wings of the building, with a lawn with relatively formal planting of grasses and perennials, enclosed with low evergreen hedging but open to the park; and second, a wilder 'secret' garden on the North side of the building, accessible only from the Gallery. Sarah Price, Chelsea Gold winner whose planting schemes at the Olympic Park in 2012 had established her as a major name, was commissioned to develop the design and planting, and a gardener was added to the permanent Gallery staff to maintain these spaces and develop biodiversity.

Through careful management and early intervention to trim roots and prepare the ground where construction would happen, only one tree was lost as a result of the project. Planting in the new gardens and green roofs on two sides of the building increased diversity, and, since re-opening a new community garden with accessible raised beds and pathways, a wildflower meadow and an outdoor classroom have been added (Figure 14.2). The Gallery has recruited a large team of garden volunteers who help support ongoing planting and maintenance. The green space around the Gallery and the engagement programmes it inspires have become a USP for the Whitworth, with the gardens offering spaces for play, relaxation and contemplation. The siting of a number of sculptures in the gardens and across the wider park has truly established the Whitworth as 'the gallery in the park'.

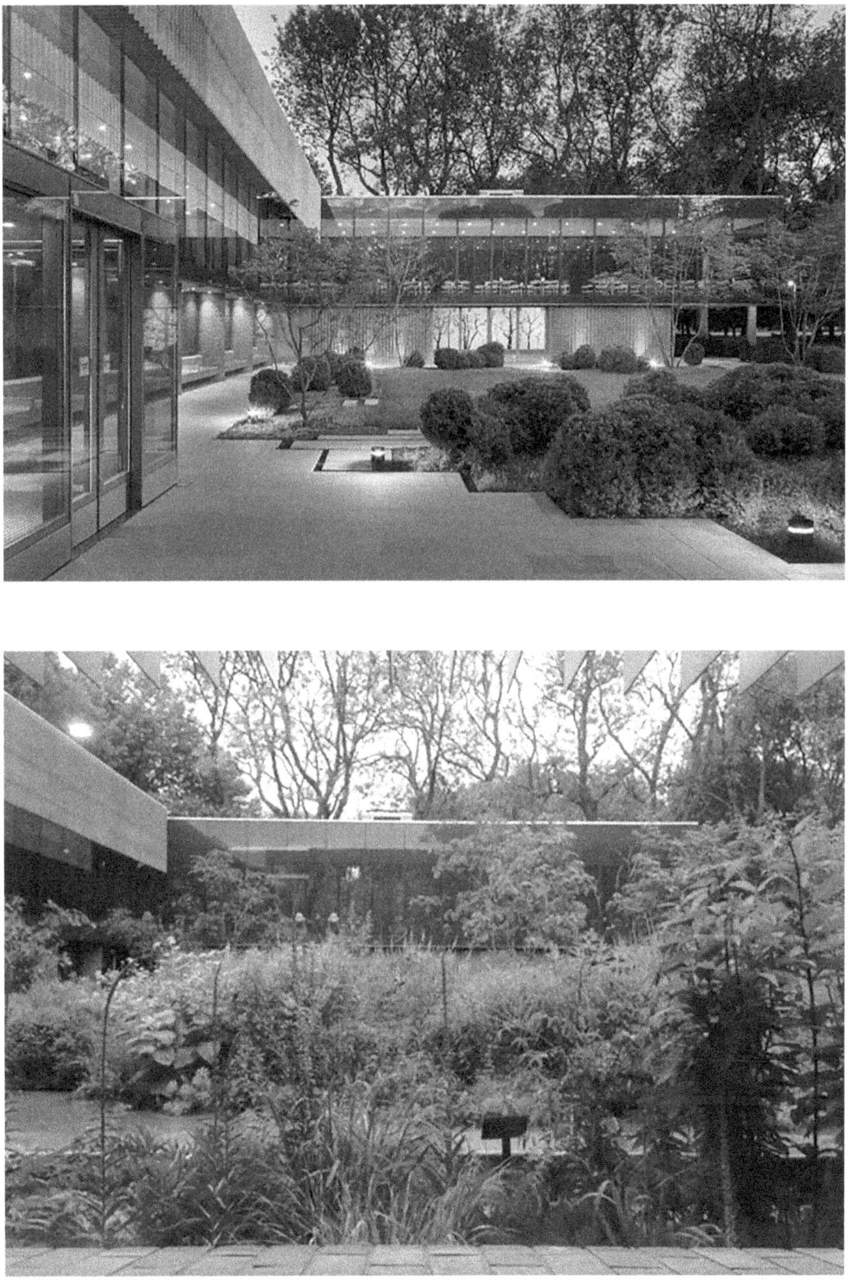

FIGURE 14.2A AND B The Art Garden in 2015 and 2017

Continued commitment

The Whitworth team delivered a major capital development project at a time when cultural budgets were decreasing. Realistically, we are unlikely to see a time when this is reversed. Creating a relatively 'architecturally self-reliant' building and avoiding high-tech solutions to day-to-day operation – from window blinds that are operated by hand to passive air control – deliver ongoing savings and, at a time when public institutions are closely scrutinised, allow the Whitworth to continue to share good practice across the sector and with visitors.

In addressing the climate crisis, museums not only have a responsibility to get their own 'houses in order'; they must also be committed to sharing the urgencies with audiences and help them to engage in thought and activity that will make a difference. Families who originate from the Global South, the part of the world most impacted by climate emergency, make up a large part of the communities around the Whitworth. Engaging these communities in the general life of the Gallery was a core aim of the capital project, and the Whitworth is committed to working to address issues of local and global social justice. As a trusted educational and social space, the Whitworth has a unique opportunity to share its environmental aspirations with these communities on its doorstep and will continue to do this through our actions – from exhibitions and engagement programmes to the offer in the café and shop, living our values in all aspects of our work.

At a time when energy bills are crippling organisations, the financial impact of over-heating/cooling buildings and the public awareness of the problem may well support our carbon-reduction targets. We now expect buildings to be cooler in the winter and warmer in the summer. The Whitworth has taken a pragmatic approach, asking staff and visitors to dress appropriately for the season. This is just the start. Programming is also taking a seasonal approach, considering Relative Humidity demands and avoiding showing exhibitions containing sensitive objects at times of year when we know large fluctuations are likely to occur. More use of the Gallery's own extensive collection offers the flexibility to programme in this way, while also reducing the need for high-cost and high-carbon transportation.

What did we learn?

It is interesting to revisit the project over ten years after the original brief was written and consider how far things have moved on in relation to the development of sustainable buildings. The check list approach to sustainable technologies used at the time already seems dated, and some of those technologies adopted for the Whitworth which seemed innovative at the time are now obvious, even regulation. This acceleration is, of course, a reaction to the increased urgency around climate emergency and demonstrates that the sector is serious in intent when it comes to developing more sustainable buildings.

It is hard to imagine embarking on this project now and what would need to be different to justify it. While the elements of the project that have proved to be the most successful for carbon reduction are the lighter touch, rather than high-tech ones, they have not entirely delivered the expected outcomes in their entirety.

The structural design of the wrap-around extension to create an environmental buffer zone and the creation of additional sources of natural light have had some impact and have certainly enhanced the feel of the building and visitor experience. Reconfiguring and rationalising the building meant that the extension required was relatively modest, increasing the floor space by a third but doubling the public space, including opening up new spaces for commercial use. This not only reduced the carbon outputs during construction, it also means that utility use has been reduced and that more space was made available for exterior landscaping. Increasingly, with flooding becoming a major issue for Manchester, the footprint of the park, set within a heavily built-up area, acts as a soak for excess water.

However, in terms of creating the right environmental conditions for collections and exhibitions, the changes have not been without challenges.

Changes to the environmental conditioning are not easy to adopt, requiring a fundamental shift in thinking across staff teams, with stakeholders, and across the sector – a genuine willingness to do things differently. While, based on projections, the Whitworth staff bought into these changes, the actual results have meant that the project has required a major re-think. As it has turned out, the thinking was rather ahead of its time. The removal of air chillers, humidifiers and de-humidifiers, replaced with new mechanical and passive ventilation, enables the Gallery to operate within the temperature and humidity bands intended, but the 10% in 24 hours rate of RH has proved more difficult to achieve. Given the nature of the Whitworth's collections, loan works and frequency of changeovers, buffering through micro-climates and air tight cases is not always possible. These unacceptable fluctuations have meant that that alternative solutions have had to be found to ensure the correct environment for the Gallery's own collection, including the temporary siting of movable humidifiers/dehumidifiers in some spaces. In particular, such measures have had to be implemented in order to secure loans from institutions still operating under traditional parameters and who have found the Whitworth's approach to the management of temperature and humidity challenging. Staff have needed to be personally committed to, and knowledgeable about, the methods being used and able to clearly articulate the approach and benefits to colleagues in lending/partner institutions.

This disappointing result is not down to incorrect projections alone. Little thought was given at the time to the rate of change in local climate, and even if it had been, it is unlikely that our predictions would have been correct. Not only has colder and wetter weather made environmental control difficult, it has also meant that the fuel bills have risen sharply – a stark reminder that making changes for environmental reasons does not necessarily result in financial savings.

In fact, the Whitworth's gas and electricity consumption has risen by around 20% post-construction, and although this is lower than the increase in the size of the building, it fails to deliver the reduction that was hoped for when the brief was prepared. Having said this, the building welcomes more visitors that it previously did and is used in a different way, so a like for like comparison is not particularly meaningful. Given the increase in fuel costs, however, this increase in consumption still represents a significant financial burden.

While discouraging to have not been able to deliver successfully against the brief at the Whitworth, the subsequent project at the Gallery's sister organization, Manchester Museum, has been able to draw on the learning from the Whitworth. Having originally planned to implement a similar system, the Museum has changed tack and implemented a more controlled system. While this is a compromise and feels, to some extent, like falling back on traditional ways of operating, it offers the opportunity to work in a low-carbon way most of the time, while also ensuring that more demanding environmental controls can be met when necessary.

On re-opening in 2023, Manchester Museum's system will be able to operate in five modes:

Mode 1 (Close Control 1): Temperature 16–25°C, RH 40–60% (no more than 10% fluctuation in 24 hours).

Mode 2 (Close Control 2): Temperature 16–25°C, RH 40–60% (no fluctuation criteria).

Mode 3a (Temperature Control): Temperature 16–25°C, no RH control.

Mode 3b (Temperature Control): 16°C, min 07:00 – 18:00 hours (for when no delicate objects are in the space).

Mode 4 (Fresh Air Override): 16°C minimum. Two-hour time period of fresh air only, no recirculation of air. This mode is designed for when dusty or vapour-heavy work is happening during builds, and the system reverts to Mode 3B after two hours.

In addition, Manchester Museum will case vulnerable objects in environmentally controlled micro-climates, allowing for broader parameters in the wider room.

It should be noted that all parties involved in the Whitworth project team were equally committed to carbon-reduction and were keen to deliver the best possible results through a lighter touch approach. Careful consideration was given to design team and contractor appointment to ensure the team fully understood the needs of the collection, of the historic building, of staff and visitors, as well as the organisation's aspirations to deliver a sustainable building. Although embedded carbon targets were not set in the brief, there was an expectation that this would be considered. Were the project to be undertaken now there would need to be a much more robust approach to this? With the University and city looking to deliver net-zero carbon by 2038, there is a realization that this cannot be achieved without robust Scope 3 carbon accounting being undertaken (see the Carbon Trust's Scope 1–3 emissions definitions). Had this approach been taken, the Whitworth's steel and glass extension may now look very different.

Conclusion and recommendations for the sector

The stark truth is that capital projects are rarely likely to come without a significant inherent carbon cost. If future carbon reduction is the goal, it may take some time for this to offset the costs of the initial construction work. With carbon-reduction technology in a constant state of development, and given the average length of capital projects, it is difficult to be sure that selected solutions will be the best ones. In fact, companies developing these technologies will be keen to experiment – which will be at your expense. Ensuring you have the right team around you is essential. The known unknowns are difficult enough to negotiate, but you will definitely need people who can help identify the unknown unknowns.

The museum and heritage sector is packed with people who live and breathe collections, buildings and visitor experiences. Architects are passionate about design. Construction companies (and their sub-contractors) rarely share this level of passion or fully understand the unique needs of museum buildings. Setting out a powerful and coherent brief is essential for a successful project, and the aspirations for carbon-reduction need to be a core part of it. Clear and achievable goals need to be set, with a full understanding of what's possible within the constraints of what is likely to be a complex, non-standard project.

As we move rapidly towards a time when the legal and ethical imperatives for carbon reduction will become tighter, for the design and construction industry and the cultural sector (especially where direct government funding is being used for capital development), there should be an expectation that things that have been standard practice in the past may no longer be possible.

Delivering challenging carbon-reduction targets that will meet the UK's 2050 targets will mean difficult decisions for all industries, and the museum sector is no exception. Large-scale capital projects will need to have an even stronger justification than ever, historic buildings will need to be modified and listed building permissions may well need to be relaxed. Decisions will have to be made regarding what goes on display, where and when. Curatorial and collection care teams will need to have better understanding of how adopting a lighter touch approach to display conditions impacts their collections and will need to develop engaging and transparent ways to communicate this to visitors.

Whatever financial and staff capacity resource has been set aside in a project to deliver sustainable solutions, it will never be enough. However, possibly more of a challenge is time. The conversations around sustainability-related changes, particularly if they will result in new ways of working, need to be discussed early in the project. Everyone's buy-in is essential, and they will need time to fully understand how the changes will impact them and to air their concerns. During the project, it is essential to ensure the plans remain current. This area of building technology is moving fast, and the time (and resources) to be able to adopt new thinking will offer long-term savings. At the end of the project, a significant amount of time needs to be set aside to stabilise environments and test new systems and to make any necessary changes prior to full operation. Whatever is done, though, it is important to accept that it won't necessarily be a long-term solution and that further changes may well be needed to adopt new technologies and respond to a climate emergency that will continue to throw up unexpected and unwelcome challenges.

Note

1 https://www.whitworth.manchester.ac.uk/visit/thenewgallery/history/

Reference

Manchester Climate Change Action Plan 2020–25. https://www.manchester.gov.uk/downloads/download/6033/mcc_climate_change_action_plan

15

RESPONDING TO THE CLIMATE CRISIS AT LEEDS MUSEUMS & GALLERIES

Lisa Broadest and Yvonne Hardman

What's the story?

Museums and galleries can be a powerful resource in addressing the climate and ecological emergency, and many cultural organisations are rising to the challenge to produce work on environmental themes, raise awareness with audiences and reduce the environmental impact of their own activities.

Leeds Museums & Galleries has galvanised this approach in a city which has a population of around 800,000 people. It is one of the largest local government-managed museum services in the UK with around 1.3 million objects and nine venues located across the city, reaching up to 1.7 million visitors a year.

When Leeds City Museum opened in 2008, the role of natural science collections in environmental messages, alongside the research data they hold, was brought to the fore in the Life on Earth gallery.

Leeds Museums & Galleries formalised this commitment to addressing environmental sustainability in 2012 with its first environmental policy. This recognised the need to ensure that environmental messaging was accurate and that it could have a significant impact on visitors' attitudes and behaviour.

By 2016, a group of colleagues came together to question how the organisation made exhibitions. Temporary in nature, they are traditionally resource-heavy and wasteful. We wanted to do things better and set new standards to roll out across the organisation. This proved to be a significant moment of change.

When Leeds City Council, our parent organisation, declared a climate emergency in March 2019, this provided a further impetus to Leeds Museums & Galleries and cultural organisations across the city to proactively respond. The Leeds Climate Commission was tasked with producing a science-based roadmap for Leeds, working with experts from the University of Leeds, which was published

DOI: 10.4324/9781003347606-17

in the same year. The Commission also worked collaboratively with Leeds City Council to co-host a city-wide conversation on the climate emergency. The Council's roadmap sets out plans to become a net zero city by 2030.

This sets the context for Leeds Museums & Galleries' work, striving to be innovative in our practice, maximising our position within our parent body for environmental benefit and sharing our learning with the sector.

Making change

Colleagues recognised that the model of non-touring temporary exhibitions in the sector has often been about single use of materials rather than focused on re-use. Some curatorial blue-sky thinking in 2015 about what collections might look like 100 years from now led to an exhibition that detailed the impact that animal behaviour has on the natural world, and what we might learn from it, which brought the two things together.

They wanted to put sustainability at the heart of exhibition-making practice as part of what became the award-winning exhibition: 'Beavers to Weavers: The Wonderful World of Animal Makers' which opened in July 2018 at Leeds City Museum. The approach was underpinned by Leeds Museums & Galleries' renewed commitment to sustainability, which was laid out in its updated Environmental Sustainability Policy and Action Plan. This recognised the need to take a more holistic approach to sustainability and the part collections and exhibitions could play in that.

The exhibition focused on the natural science collection and celebrated the beautiful things made by animals; how they build their homes, make tools to catch food or give gifts to attract their mates, using resources only according to their need. Through the interpretation, we wanted to encourage our visitors to think how they used resources, and the impact they had on the natural world. As such it seemed remiss to not seek to make the exhibition in as sustainable a way as possible and embed this into every step of the exhibition-making process.

In order to take this forward, we devised our own set of environmental principles for the project, which at the time were quite ambitious, undertaking all development work with these in mind and constantly evaluating our plans against them.

The eight principles focused on the following:

- Object loans and transport: working with lenders to encourage them to re-use existing crates and packing materials or determine whether they could be reused after the exhibition or whether they could be 100% recyclable; whether loans could be transported using public transport (two lenders agreed to this) or through greener transportation such as using electric vehicles.
- Environmental conditions: using air handling systems to maintain stable environmental conditions is one of our main energy uses particularly at Leeds City Museum. The aim was to reduce the requirement for constant monitoring, but we needed to work with the lenders and with the wider Leeds Museums & Galleries' team to ensure the objects would not be put at risk.

- Design and build: sustainability criteria were written into the design brief and asked the designers to think differently and assess what materials and resources were already available and in existence before commissioning anything new. We wanted to procure more sustainable exhibition graphics which considered all elements of production from the materials they were printed on, to the ink that was used, as well as securing them from a local provider to minimise travel. With regard to the build, the focus was on reusing display cases and plinths, working with recycled paint from a local community enterprise and reusing all other materials where possible.
- Interpretation: the approach focused on engaging visitors to think about how they use resources in their own lives and the impact that this has on the natural world. In terms of materials, we wanted to find a replacement to foamboard. We ended up using a corrugated cardboard material made from a 100% recycled honeycomb core, faced with Forestry Stewardship Council (FSC) certified material which is suitable for direct printing and 100% recyclable after use. It is now in frequent use in the sector as venues strive to move away from plastics.
- Community engagement: part of the interpretation of the exhibition was going to use installations to demonstrate some aspects of animal behaviour to visitors. The project anticipated that rather than commissioning these from external companies, they would be created in conjunction with communities in the city, building an audience base for the exhibition who would advocate for living more sustainably.
- Marketing: the project specified that all printed promotional material should come from FSC-accredited, recycled sources and be 100% recyclable. It was agreed that consideration should also be given to boosting digital streams and reducing physical print (at a time when more printed material was still being created in the sector).
- In-gallery activities: the aim, in line with overall exhibition design, was to utilise reused and recyclable/reusable activities in the exhibition gallery and determine how things could be left in situ to be then reused by other visitors. We made our own paper from scrap paper generated by the museum service to use in craft activities and also sourced materials from a local social enterprise focused on the re-use of scrap materials.
- Learning programme: this took the same approach as in-gallery activities. The team sought to develop activities for informal learning (e.g. during the school holidays), which would use low-impact resources and which could be reused or recycled after they had been created. The project wanted to look at how the activities could deliver a message about sustainability and care for the environment to visitors.

The exhibition welcomed over 48,000 visitors and won the Museums Association's Museums Change Lives judges' award for environmental sustainability in 2019.

Next steps

The learning from the 'Beavers to Weavers' exhibition was taken and rolled out across programming, and a commitment was made by Leeds Museums & Galleries and Leeds City Council to:

- Reduce the environmental impact of transport and travel and promote suitable alternatives
- Reduce the use of material resources and improve environmental quality
- Protect, enhance and increase biodiversity

From the sustainability principles, we saw the development of the exhibition 'Fast × Slow Fashion: Shopping for Clothes in Leeds 1720–2020' again at Leeds City Museum. The exhibition explored the relationship between clothes, shopping and people in Leeds. It examined ready-to-wear clothing and bespoke garments recognising the consumption of mass-produced clothing, which is widely available and seen as more affordable. The difference here was that the exhibition recognised that as a population, we are facing the global climate crisis and was more direct in its messaging. Feedback did however recognise that this could have gone further and is a useful gauge for us as an organisation as to how strong the interpretation can be around this type of messaging. We are very mindful that we are working in a city where people are living in poverty and didn't want the messaging to make anyone feel excluded. Public programming for the exhibition brought a focus on environmental sustainability including a conference with the University of Leeds around fashion consumption, upcycling workshops and, during the first UK-wide COVID-19 lockdown when the exhibition became part of our virtual visit, we linked with online tutorials being offered by a local vintage store. The exhibition has led to our involvement in other projects with organisations such as Zero Waste Leeds, whose aim is to help Leeds become a zero-waste city by 2030.

Another exhibition project building on the principles and developed against the backdrop of the pandemic was Natural Encounters at Leeds Art Gallery. This exhibition explored the many different strategies artists have used to approach, interpret and respond to nature, and it explored the important role that art plays as a tool for reflecting on current ecological issues and our ongoing relationship as a species with the natural world we inhabit. Natural Encounters was shortlisted for the UK's Museum & Heritage Awards environmental project of the year.

A new perspective on the important role that art plays in relation to nature and wider sustainability issues also came from the context of the first national COVID lockdown. It gave rise to reflections on how our relationship with nature was transformed on multiple levels – from the lack of traffic noise and aeroplanes in the skies, and the return of audible bird song, to a growing appreciation of the green spaces around us and a closer experience of nature changing with the seasons. In particular, it reminded us that access to nature, whether in gardens, public parks or wilder spaces, is critical to our mental and physical well-being and that this is something that needs protecting.

A number of colleagues volunteered, via Leeds City Council's Woodland Creation Scheme, to plant trees over the course of the exhibition to help reach the Council's aims to plant 5.8 million trees, contributing towards making the city carbon net-zero by 2030. One strong element of the public programme linked with Leeds Art Gallery's Youth Collective and encouraged young people to work with a variety of climate crisis youth work organisations. The accompanying public programme, hosted online due to the pandemic, addressed the exhibition's urgent themes and worked to bring audiences together in those difficult times. 'Lockdown Patchwork', a music commission by Leafcutter John made with crowd-sourced nature sounds was launched a year on from the initial lockdown as part of the exhibition in partnership with Huddersfield Contemporary Music Festival.

Practically, Natural Encounters wholly reused existing plinths, labels were printed on recycled eco-board and 95% of the works in the exhibition came from Leeds Art Gallery's collection while the small number of loans travelled on part-loads with other consignments or via Leeds Museums & Galleries' electric van, continuing to embed the sustainability principles.

Leeds Museums & Galleries' most recent project during summer 2022 has been to apply the learning from making temporary exhibitions in a more sustainable way to refurbishing a major permanent gallery for the first time. This had been done on a smaller scale at Leeds Industrial Museum in 2019 with the re-fit of the Power House gallery, which followed the environmental principles and brought a more critical messaging to the subject of engines and how they impact our world.

The Life on Earth gallery at Leeds City Museum showcases the natural science collections which are formally designated by Arts Council England as being of national importance. The gallery originally opened with the rest of the museum in 2008 and hadn't seen a major overhaul beyond updates to specific display cases or changing specimens on display for preventative conservation reasons. Although the gallery's interpretation had always had an environmental message, it was felt that this could be stronger to recognise the climate and ecological crisis and provide a call to action for audiences in their own lives. At the same time, the interpretation was to address colonial legacies within the natural science collection.

In embedding the principles developed for exhibition making, a key aspect was to appoint a design company who were able to share in this ambition to refurbish an existing gallery in the lowest carbon way possible, and this was written large into the tendering process. We were able to appoint designers, Thomas Matthews Communication Design,[1] who both had a track record of working in this way and proactively wanted to meet the challenge of finding new materials that were appropriate to the project, with an intentionally longer life span than some of the recyclable materials used in temporary exhibitions, such as cardboard for labels and panels. We were also faced with questions we hadn't previously considered, for example were we comfortable with using recycled plastics or did we want to specify non-plastic materials? Some solutions ended up being pragmatic ones; for example some newer recyclable materials for interpretation inside display cases had not previously been Oddy-tested[2], and the tight timescales of the project meant that we had to choose something that had already been tested. We worked with the existing infrastructure of the gallery in terms of display cases, having had the case seals serviced.

Life on Earth displays animal taxidermy from around the world and the UK in statement display cases alongside other objects, and this physical exhibition infrastructure was maintained. Objects were placed to make new connections between different animals, reflected in the interpretative narrative. The revised interpretation presents information about evolution, ecosystems, biodiversity, the impact of humans on nature, extinction and conservation. The provenance of the collection is brought to the fore, and the interpretation is explicit about colonial legacies. The gallery's interpretative approach is to make clear, factual statements about how the climate and ecological crises impact on both a global and local level and not seek to soften the messaging of how serious the situation is. At the same time, the tone is encouraging in terms of how visitors can make a difference through their own actions. This quote from the primatologist Jane Goodall, which appears in the gallery, sums up the approach, 'You cannot get through a single day without having an impact on the world around you. What you do makes a difference, and you have to decide what kind of a difference you want to make'.

As part of the refurbishment, we commissioned research to undertake a carbon impact assessment with the aim of establishing a best practice baseline of carbon emissions from this type of retro-fit. The project also developed a carbon tool that will provide a benchmark assessment of the lifecycle carbon footprint of the development, production and hosting for future exhibitions (both permanent and temporary), as well as useful design guidelines, RAG (Red, Amber, Green) rating for materials and a decision-making tree to improve sustainability. This work was undertaken by Useful Projects,[3] which is part of the same group of practices as the gallery designers. Two members of the team working on this project also worked as part of the URGE collective as advisors on the Design Museum's exhibition 'Waste Age: What Can Design Do?' which opened in October 2021.[4]

The scope of the initial assessment included materials, transport, travel by the consultants as part of their work, communication and digital, production and utilities. As the original gallery fit out took place in 2008, there was insufficient data to calculate the original impact so a 'business as usual' scenario was developed and benchmarked against actual delivery. Data was matched with carbon conversion factors from the most appropriate available resources, consisting of recognised life-cycle inventory databases, Environmental Product Declarations (EPDs), academic research as well as industry publications. Where possible and appropriate, factors were adapted for the UK context. For recycled and reused materials, a factor of 0.5 was applied to allow for the embodied carbon to be allocated across at least one previous application. Knowledge of the design process was crucial for this, and where materials were prevented from going to waste (i.e. off-cuts), their carbon footprint was considered negative due to them avoiding emissions from incineration or landfilling.

The results from the assessment confirmed that the total net carbon footprint of the retrofit was approximately 2,300 kg CO_2e, around the same as a one-way premium economy flight from the UK to Hong Kong. A saving of approximately 1,530 kg CO_2e was calculated. Given that sustainability was part of the original brief, a saving was anticipated. This was achievable as the appointed designers assessed and reused some existing components, and through their decision-making selected lower-impact materials compared to the 'business as usual' exhibition design. These decisions equated to a reduction of design-related emissions of roughly 55% and an overall exhibition footprint reduction of 40%.

One of the largest shares of emissions at 47.5% was apportioned to production materials including relatively small quantities of petroleum-based materials such as vinyl products and significantly larger quantities of composite or non-petroleum-based materials such as wallpaper and recycled MDF. Sixty-nine per cent of the savings had been made through the reuse of large structural steel panels and re-wrapping them and a further 28% by selecting products made from recycled resources. Additional savings were also made by preventing 31 m^2 of Perspex off-cuts going to waste and reusing them in this installation. Designing with purpose-made Perspex would have contributed an additional 722 kg CO_2e. The other largest share of emissions at 43% related to the electricity and district heat required to operate the gallery space during the retrofit installation period. It was further recognised that while the museum receives its heating through a district heating network, this method still incurs carbon emissions.

A number of observations and recommendations have been made following the outcome of this project, which include building in longer timeframes in order to identify appropriate materials; determining whether materials are categorised as reused, reclaimed or recycled; knowing in more detail what we already have with a list going out with every new brief and setting a target for re-use, to build a list of preferred local suppliers that are working to recycle and reuse materials and to build in the carbon tool as a contractor prerequisite developed as part of this project. This tool will ensure materials are assessed early in the design process to inform decisions with an order of priority ensuring materials are chosen as to whether they are reused, recycled or made from renewable materials. Composite, petroleum-based and single-use or difficult-to-recycle materials should be avoided. We hope to use this project as an exemplar of how to approach the refurbishment of other permanent galleries at Leeds City Museum and other Leeds Museums & Galleries' sites, as well as sharing learning with the sector.

Over the past two years, as part of this sharing process, we have worked with partner organisations in Leeds' twin city of Lille on a knowledge exchange focused on the climate crisis and how we are responding. Funded through the British Embassy in Paris, the Sustainable Culture in Leeds and Lille (SCILL[5]) project provides the opportunity for a range of cultural organisations to explore different aspects of sustainable culture, exchange ideas with peers and importantly build new international relationships. The long-term aim of the project is to reduce the carbon footprint of the cultural investment both cities make through the exchange of best practice and action planning. This has been a valuable experience of shared learning, particularly for Leeds Museums & Galleries with colleagues at the Palais des Beaux-Arts de Lille, who have also been taking action to make exhibitions in a more sustainable way and reduce carbon footprints.

Next steps for Leeds Museums & Galleries in this area of work are looking at setting specific targets for reusing materials in exhibition production and minimising waste. We are considering introducing a carbon budget for each exhibition shaped by our experience with Life on Earth and the partnership with colleagues in Lille, who have been working in this way. We will continue to review the necessity of any loans (both in and out) and to work with lenders to minimise the carbon footprint of any loan. Sustainability will always be written into design briefs now shaped by the observations and recommendations from the Life on Earth project. We will continue to follow the development of new materials and be alert to the opportunities these may present.

What about the buildings?

Leeds Museums & Galleries' nine venues are primarily historically significant sites. They are among the city's finest civic buildings, including Temple Newsam, one of the great houses of England, bursting with treasures and tales and over 500 years of history; Leeds Art Gallery which opened in 1888; Leeds City Museum which was built in 1865–1868 as a mechanics' institute and Leeds Industrial Museum, which was once the world's largest woollen mill. Each building provides its own challenges in relation to energy use and efficiency.

Progress towards reducing carbon has come directly from Leeds Museums & Galleries as well as benefitting from being part of Leeds City Council. This has come with some foresight, for example the museum service's only fully modern site, Leeds Discovery Centre, a collections store and research centre, opened in 2007 and incorporates solar panels on the roof to contribute to the building's energy use. Changing to LED lighting has been happening over a number of years, instigated by Leeds Museums & Galleries and latterly picked up for Leeds City Museum as part of the Council's Public Sector Decarbonisation scheme. This is funding from the UK government's Department for Business, Energy and Industrial Strategy and has enabled a major programme to install £25.3 million of energy efficiency upgrades across 40 public buildings in the city. It is expected to have reduced the city's carbon footprint by more than 4,000 tonnes each year. Benefits directly to Leeds Museums & Galleries are via connecting Leeds City Museum and Leeds Art Gallery to the Leeds PIPES network, upgrading the building management systems at each venue and upgrading all lighting at the Museum (changes were made at

Leeds Art Gallery in 2017). The Leeds PIPES network is a district heating network providing low-carbon heat and hot water, reusing heat which is already being produced at Leeds's Recycling and Energy Recovery Facility. Almost 2,000 homes and numerous businesses and organisations in the city are being connected with the aim of saving over 16,000 tonnes of carbon annually.[6]

Adaptation is another important aspect of the city's approach impacting positively at Leeds Museums & Galleries. Two of the organisation's museum sites were affected by significant flooding on 26 December 2015 and in February 2019: Thwaite Watermill and Leeds Industrial Museum. The Boxing Day 2015 storm caused the city's numerous waterways to converge to flood homes and businesses at unprecedented levels with long-term implications. At Leeds Industrial Museum, the height of flood was above the level of the 'Great Flood' of 1866, when a heavy goods train from Bradford to Leeds plunged into the River Aire from a viaduct that had been weakened by the severity of the flooding. Both this historic event and the 2015 floods are commemorated with metal plaques positioned on one of the museum's buildings according to the flood height; the waters in the more recent event reached around 2.5 m which was approximately three times higher than the 1866 flood. It resulted in the museum being closed to the public for three months for the clean-up work and to re-connect power to the site, which had been lost. Significant damage was also seen at Thwaite Watermill, which is essentially nestled on an island between the Leeds and Liverpool Canal and the River Aire. For both sites historically, water was essential to their functions, but today it provides a challenge to their preservation and use as heritage museum sites.

Part of the adaptive response has seen Leeds City Council and the Environment Agency working in partnership with appointed contractors to implement the city's Flood Alleviation Scheme, recognising that climate breakdown means that extreme weather events are happening more often, and therefore preparation is increasingly important. Close to Thwaite Watermill, one of two moveable weirs was installed in the River Aire in 2017 in what was a UK first. The moveable weirs can be lowered to control river levels and create more room in the channel when required during periods of heavy rain that would previously have resulted in flooding. The benefit of the weir has been seen in the wider area when in use. This enables the water to be dispersed more quickly, although the museum has still seen some flooding during these events. At Leeds Industrial Museum, a flood wall is being built around perimeter of the site as part of phase two of the Flood Alleviation Scheme along the River Aire. The scheme was originally designed to provide a one in 200-year level of protection against flooding, but due to changing weather patterns, this may change to a one in 100-year level.

More broadly, within Leeds Museums & Galleries, in 2019, the service appointed its first green champion, within the role of the Head of Operations & Enterprise, who is tasked with not only leading improvements at our venues but also importantly in embedding an organisational culture of environmental awareness and knowledge in what is and has to be a collective endeavour. At the start, inspiration came from the work of economist Kate Raworth, including her publication *Doughnut Economics*[7] and seeing her speak a number of times in person and online, including the President's Lecture of the RSA in 2020.[8] Subsequent follow-up research around the circular economy included helpful webinars by Julie's Bicycle[9] and the implementation of the Leeds Doughnut by Climate Action Leeds.[10] Work has also been influenced and driven by the Leeds Climate Commission roadmap.[11]

The post is also pivotal in our partnership work in this regard and sits on the project board for the Public Sector Decarbonisation scheme, on the National Museum Directors' Council Environment and Ecology working-group and as a steering group member for Sustainable Arts in Leeds, known as SAIL.[12] In early 2018, a group of organisations and companies from the cultural and creative sectors of Leeds came together to discuss how we could respond to the climate crisis, and SAIL was the result. It brings together organisations, individuals and networks, with an objective to speak and act with our collective strength, using one voice and ensure that the cultural and creative sectors in the city meet the net-zero target for 2030. SAIL provide carbon-literacy training for organisations working in partnership with the Carbon Literacy Project. In December 2020, colleagues from across Leeds Museums & Galleries' nine sites and all departments took part, and individuals became accredited as carbon literate. The Carbon Literacy Project states that carbon literacy is having 'an awareness of the carbon dioxide costs and impacts of everyday activities, and the ability and motivation to reduce emissions, on an individual, community and organisational basis'.[13] SAIL state that a study by a sustainability consultancy firm on the effectiveness of Carbon Literacy training. Discovered there is between a 5 and 15% carbon saving per person for those that undertake the training' and 'a study at the University of Leeds which Opera North and ITV took part in also found that Carbon Literacy is an effective way of instigating lasting change.[14]

As an organisation in receipt of regular funding from Arts Council England since 2012, this has provided additional focus to embed environmentally driven policies and planning within Leeds Museums & Galleries. At this point 10 years ago, Arts Council England became the first cultural body to make environmental reporting and policy part of its funding agreements and launched a partnership with Julie's Bicycle, a pioneering not-for-profit organisation mobilising the arts and culture to take action on the climate and ecological crisis. Annual reporting of energy use, waste and recycling data has provided an overview and a way of tracking progress through an online dashboard. Joining Julie's Bicycle's Spotlight programme in 2019 furthered this work by proactively setting environmental impact reduction objectives for building energy use as well as developing good energy management practices. LMG has utilised the Creative Green Tools developed by Julie's Bicycle as part of our reporting as a National Portfolio Organisation.[15] Leeds Museums & Galleries was also part of the Happy Museum's No Going Back programme,[16] which supported the development of thinking about the Life on Earth refurbishment. Work during 2021–2022 on a new strategic plan for LMG gave the opportunity to reflect on the UN Sustainable Development Goals within the work of LMG.[17]

Other areas of focus have been on increasing rates of recycling and reducing waste, including in cafés and retail. Tracking this in comparable quantitative terms remains difficult as contracts change as different organisations work differently within this area. Leeds City Council has led on the removal of single-use plastics where possible. We have worked with a charity in the city to reduce food waste, and we are working with University of Leeds to put food miles on menus. Sustainability has now been given weight within Leeds City Council's procurement policies. This approach links back to our work on making exhibitions and refurbishing galleries in ways which embed the reuse of materials from the outset.

We have expanded our community gardening projects, building on the success of the Colour Garden at Leeds Industrial Museum, so called because of its focus on growing plants traditionally used in the dyeing of yarn and fabric. We have revived and added to raised beds at Kirkstall Abbey, the ruins of a Cistercian monastery, dating from 1152. The growing area at the site was originally developed as part of a Happy Museum project almost ten years ago and takes inspiration from the lives of the monks who once called it their home and would have had an ethos of self-sufficiency. For each project, we work in partnership with a community-gardening organisation in the city, Hyde Park Source. They embed the five ways to well-being[18] and the 12 principles of permaculture[19] in all their work. Our collaborative gardening projects enhance existing green spaces, increase biodiversity and improve the visitor experience. Most of all, they focus on the well-being of participants and exemplify the connection between nature and supporting mental health.

How can we summarise what we have learnt?

Over the last ten years, Leeds Museums & Galleries has seen a substantial period of growth, but we are aware that we are living and working in a changing world. Priorities and in turn expectations are changing with an even greater focus on relevance and inclusion, environmental responsibility, financial sustainability, health and well-being and digital reach.

We have a strong organisational commitment to responding to the climate and ecological crisis. Our parent body, Leeds City Council, declaring a climate emergency before the UK government gave Leeds Museums & Galleries a strong mandate to take this work forward.

Alongside that internal commitment is the importance of making outward connections, both with audiences and with partners. Listening to audiences and working collaboratively in partnerships with organisations who can offer mutual support are crucial, as is the challenge of keeping up to date in our thinking, our training for colleagues, in our use of materials and resources.

There's always more to do, but we have come a long way in the last decade.

Notes

1 'TM Is Ready', Thomas Matthews Communication Design, accessed 8 January 2023, https://thomasmatthews.com/tm-is/
2 Oddy testing was developed at the British Museum in 1973 by conservation scientist William Oddy. It tests whether a material is appropriate for use in an enclosed space such as a display case with collection objects.
3 'Useful Projects Homepage', Useful Projects, accessed 21 December 2022, https://usefulprojects.co.uk/
4 Working to make change in 'Waste Age', Design Museum, accessed 21 December 2022, https://designmuseum.org/exhibitions/waste-age-what-can-design-do/working-to-make-change-in-waste-age
5 'Sustainable Culture in Leeds and Lille', Global Leeds, published 27 October 2022, https://www.globalleeds.com/post/sustainable-culture-in-leeds-and-lille
6 'Council's Building Decarbonisation Programme Hits Major Milestone With More Than Half of Upgrades Complete', Leeds City Council, published 25 March 2022, https://news.leeds.gov.uk/news/councils-building-decarbonisation-programme-hits-major-milestone-with-more-than-half-of-upgrades-complete
7 Kate Raworth, 'What on Earth Is the Doughnut?. . .', Kate Raworth: Exploring Doughnut Economics, accessed 21 December 2022, https://www.kateraworth.com/doughnut/
8 RSA Events, 'When the Doughnut Meets the City', YouTube, accessed 21 December 2022, https://www.youtube.com/watch?v=GJqhmr0K-4M
9 'Watch Again: Re-Thinking Materials and Circular Economy', Julie's Bicycle, published 1 October 2020, https://juliesbicycle.com/resource/watch-again-re-thinking-materials-and-circular-economy/
10 'Welcome to the Leeds Doughnut', Climate Action Leeds, accessed 21 December 2022, https://www.climateactionleeds.org.uk/leedsdoughnut#:~:text=The%20Leeds%20Doughnut%20is%20a,up%20to%20its%20global%20responsibilities
11 Gouldson, A., Sudmant, A., Duncan, A. and Williamson, R. (2020), A Net-Zero Carbon Roadmap for Leeds, Leeds Climate Commission/ Place-Based Climate Action Network, accessed 21 December 2022, https://leedsclimate.org.uk/leeds-carbon-roadmap
12 'Doing Nothing is Not an Option', Sustainable Arts in Leeds, accessed 21 December 2022, https://wearesail.org/aboutus
13 The Carbon Literacy Project, accessed 21 December 2022, https://carbonliteracy.com/
14 'Carbon Literacy Training', Sustainable Arts in Leeds, accessed 21 December 2022, https://wearesail.org/carbon-literacy-training/
15 'Creative Green Tools', Julies' Bicycle, accessed 21 December 2022, https://juliesbicycle.com/our-work/creative-green/creative-green-tools/
16 'Footfall to Footprint – No Going Back Conversation', The Happy Museum, accessed 21 December 2022, https://happymuseumproject.org/footfall-to-footprint-no-going-back-conversation/
17 'Sustainable Development Goals', United Nations Foundation, accessed 21 December 2022, https://unfoundation.org/what-we-do/issues/sustainable-development-goals/?gclid=Cj0KCQiA-oqdBhDfARIsAO0TrGHSUCXDNTjE3YlBpHgU0_vsISt7AzMfDJxwOE6oUCLmUdftAt28W9MaAsGvEALw_wcB
18 'HP Source Principles and Pathways', Hyde Park Source, accessed 21 December 2022, https://www.hydeparksource.org/principles-and-pathways.html
19 'We Are Passionate about Permaculture!', Hyde Park Source, accessed 21 December 2022, https://www.hydeparksource.org/permaculture.html

16

SHAPING A POSITIVE FUTURE AT THE HORNIMAN MUSEUM AND GARDENS

Carole Destre and Nick Merriman

The Horniman Museum and Gardens lies in the outer part of London, UK, in a neighbourhood where 45% of the population is ethnically diverse. Its collections comprise global anthropology (including one of the best collections of musical instruments in the world) and global natural history. It also has an aquarium, butterfly house, animal walk and half-mile nature trail in 16.5 acres (6.7 ha) of gardens, including views of central London. It is London's only museum where nature and culture can be seen side by side. It has a predominantly family audience (75% of visitors) who are regular attendees, visiting often multiple times a year, over generations.

For decades, the Horniman has had a strong engagement with the natural environment. Its aquarium is over a century old, and its renowned Schools Team has worked to engage local children with the environment since at least the 1950s. In 1973, it took over the management of London's oldest nature trail, part of an abandoned railway line. From the mid-1990s, the Horniman worked with local biodiversity partnerships, and in 1996, it opened the Centre for Understanding the Environment (CUE), a grass-roofed building made of sustainable materials as a hands-on centre for learning about nature.

In 2013, responding to the catastrophic bleaching of coral reefs as a result of climate breakdown, its aquarium became the first in the world to successfully reproduce broadcast coral in captivity. Then in 2015, nine corals, transplanted from Australia's Great Barrier Reef, spawned nearly 130,000 eggs at the Horniman, resulting in a UK first – the successful in vitro fertilisation of captive corals. In the lab, it has been possible to induce corals to breed six times a year, rather than once a year in the wild, thereby accelerating selective breeding of corals that are more tolerant of warmer waters. Working with partners at the University of Derby and Florida Aquarium, these techniques are being refined

DOI: 10.4324/9781003347606-18

with the hope eventually of being able to repopulate bleached reefs with these more resistant corals.

We know from talking to our audience that the parents, grandparents and carers of the young children they bring are worried about the future that they will grow up in, as we respond to – or fail to meet – the challenges of the climate and ecological crisis. But they also report feeling overwhelmed by the issues, with their efforts to recycle their rubbish feeling rather insubstantial in the face of the melting ice sheets or the burning forests. There is a real appetite, we concluded, for positive and optimistic approaches to the crisis, which would receive a ready reception among our visitors. This commitment was reflected in a new mission for the Horniman articulated in 2018:

> *The Horniman connects us all with global cultures and the natural environment, encouraging us to shape a positive future for the world we all share.*

In turn, this was followed up by the Horniman declaring a climate emergency in July 2019 on Earth Overshoot Day – the day on which the world's population has used more from nature that year than our planet can renew in a whole year. This declaration committed us to placing ecology and climate issues at the heart of what we do, and six months later, we followed this up with our Climate and Ecology Manifesto and Action Plan. The Manifesto committed us to a series of actions including:

- an ambitious Nature + Love project to redisplay the natural history and living collections to highlight climate and ecological issues to a wider audience
- reducing waste and pollution, increasing biodiversity on our site and becoming Greenhouse Gas (GHG) neutral by 2040
- creating an Environmental Champions Community to inspire and support visitors to make their own changes
- appointing a Climate and Biodiversity Coordinator, a new role dedicated to making the changes happen

The global COVID pandemic, coming just two months after our Manifesto, caused us to reflect on our priorities and accelerate actions through a Reset Agenda, which among other things highlighted the need to double down on the climate and ecology crisis.

We formed a Climate and Ecology Action Group, chaired by the Chief Executive, to oversee the achievement and monitoring of a Climate and Ecology Action Plan, which is revised annually. In developing it we were influenced by the existing plans of the Australian Museum (2019) and the Victoria and Albert Museum (2021). The action plan runs to 2040 (when we aim to achieve greenhouse gas neutrality) and identifies 15 areas where action is needed:

• Energy	Capital Programme	Building management
• Building adaptation & climate mitigation	Air Quality	Water
• Waste	Biodiversity	Procurement and banking
• Transport	Digital	Workforce engagement
• Off-setting/carbon positive	Sector communication	Public engagement

Each of the areas has an overall goal (e.g. 'Develop and implement a plan to reduce water usage and recycle water on our sites') and a series of detailed actions attached to it, with timelines and responsibilities.

It is clear that in terms of the 'Inside-Outside' Model of climate action expounded by Douglas Worts in this volume, the great majority of actions are 'inside', about making the Horniman an exemplar of good practice in reducing GHG emissions, waste and pollution and increasing biodiversity. The 'outside' actions covered in the section on public engagement are dealt with separately through our exhibitions, events and digital programmes and will be discussed later in the chapter.

We will now set out some of the key insights we have gained in developing this programme.

What do we mean?

We made a bold declaration of an intention to be GHG neutral by 2040, without really considering in detail what this actually means, or how we would get there. We were clear that neutrality in terms of carbon dioxide wasn't sufficient, as other greenhouse gases such as methane were even more harmful. But what does GHG neutrality actually mean, did we mean Scope 1, 2 and 3 emissions or just Scope 1 and 2?

Net-zero GHG and GHG neutrality are slightly different. Net zero tends to cover Scope 1 and 2 emissions and means that the amounts of greenhouse gases that are removed from the atmosphere are equal to those produced by our activity. As it is difficult to reduce emissions to zero, residual emissions would be removed through sequestration from the atmosphere.

GHG neutrality also offsets residual emissions but has a wider definition of off-setting methods. It also encourages the inclusion of Scope 3 emissions if possible.

Scope 1 covers our own direct emissions, principally by burning natural gas on site, as well as petrol from museum vehicles, while Scope 2 covers indirect emissions from the generation of electricity we purchase. Scope 3 covers everything else, including business travel and travel by our visitors to our sites. Whether to include visitor travel in an organisation's GHG footprint is a vitally important question, as by most estimates, it comprises most of the emissions attributable to a cultural organisation. For example if it is included, 92% of overall emissions for Tate Galleries would be from visitor travel (Julie's Bicycle 2019), and it has been estimated as comprising 74% of emissions for the art world as a whole (Julie's Bicycle 2021). However, visitor travel is largely outside the control of individual organisations, and arguably if the Horniman did not exist, visitors would travel to other destinations. The issue of how to account for visitor travel in decarbonisation strategies is still unresolved (WTTC 2021) so, after some discussion, we decided to aim for neutrality on Scopes 1 and 2 and reduce Scope 3 as much as possible by implementing a Green Travel Plan, which includes for example electrifying our fleet vehicles, opting for more sustainable travel by public transport and by encouraging visitors to do the same.

The more we investigated these issues, the more murky it became. Many organisations are aiming to achieve neutrality by offsetting residual emissions through participating in commercial schemes which generally operate by planting trees. These are controversial in themselves, having been accused of 'greenwashing' emissions and harming the environment (Hodgson and Nauman 2021). There are, however, carbon-positive schemes such as that operated by the Association for Coastal Ecosystem Services[1] which helps people in Kenya protect their mangrove forests and use money from the sale of carbon credits for water, education and health.

Given this, at the Horniman, we decided to aim to get as close as possible to absolute zero emissions, while recognising that at present, we will not achieve complete neutrality by 2040, with an estimated 36 tonnes of CO_2e remaining (see report later in the chapter). We will investigate ethical carbon positive or offsetting schemes to be used in due course to manage this residue of emissions.

Data challenges

Clearly, when it comes to reductions in greenhouse gas emissions and waste and improvements in biodiversity, it is essential to have baseline data with which to measure progress. This has proved rather more difficult than we had imagined.

Through our metering of gas and electricity, we are able to convert consumption in KWh to equivalents of carbon dioxide and other greenhouse gases.[2] It is slightly complicated as the electricity grid becomes more renewable, meaning that conversion factors change over time. Metering of water should be easy, but on our site, faulty meters have meant that we have only recently had reliable data.

Calculating waste has proved particularly difficult. When we developed our action plan, our waste contractor did not provide a service to weigh waste, so it was not until we changed supplier that we were able to receive information on the volume and weight of waste produced on site and the percentage of recycling.

Business travel data has also been difficult to calculate as it requires staff to fill in forms on their mode of transport the miles travelled and then converting this to emissions.

Summary of principal targets by 2040

Action area	*Benchmark (2018–2019 unless stated)*	*Target by 2040*
Energy	528 tonnes CO_2e	As close as possible to absolute GHG neutrality by 2040 through average 5% annual reduction of 2018/2019 baseline
Water	7,172 cu m (2017–18) (7,172,000 litres); 200,000 litres recycled	6000 cu m; 500,000 litres recycled
Waste	88.08 tonnes (2020–21); 27.8 tonnes recycled (31.5 %)	60 tonnes; 30 tonnes recycled (50%)
Staff travel	25% travel to work by petrol/ diesel car or motorbike (2020 staff survey)	No staff travel to work by petrol/ diesel car or motorbike
Business travel	Benchmark to come for 2022–2023	No UK flights powered by aviation fuel; revised target
Biodiversity	Consultant report in 2022–2023 showed Defra Metric Biodiversity Calculator score for area-based habitats of 13.70 units and for linear-based habitats (hedgerows) of 1.70 units	Target by 2026, following Nature + Love project, is a biodiversity net gain of approximately 30% for area-based habitats (17.87 units) and a 76% net gain for hedgerows (3.00 units)

Practical challenges of decarbonisation

Getting down from the benchmark of 528 tonnes CO_2e to as close to 0 will clearly be the biggest task facing us over the coming years. To help us understand how we might achieve it, we commissioned consultants Harley Haddow to develop a 'road map' to GHG neutrality by 2040.

Figure 16.1 shows a visual summary of their findings. Because of the gradual decarbonisation of the electricity grid, GHG emissions would reduce to 372 tonnes by 2040 with no action by ourselves, but the only way of significantly reducing these further is to transition away from our reliance on natural gas for heating and hot water, and to ensure our energy efficiency improves through better thermal performance of our estate. The chart plots out a series of capital projects that are either planned or need to be undertaken to do this. Each of them involves an existing building or gallery and consists of replacement of gas heating by renewable energy sources such as solar panels or heat pumps and significant upgrading of insulation and related measures. Most of them are projects that will be combined with necessary maintenance or upgrading of services.

The Nature + Love project is the first of these and consists of the redisplay of the Natural History Gallery and the development of two new outdoor elements: the transformation of the plant nursery into a Sustainable Gardening Zone and a new nature-themed outdoor play area and café for children. The new café and

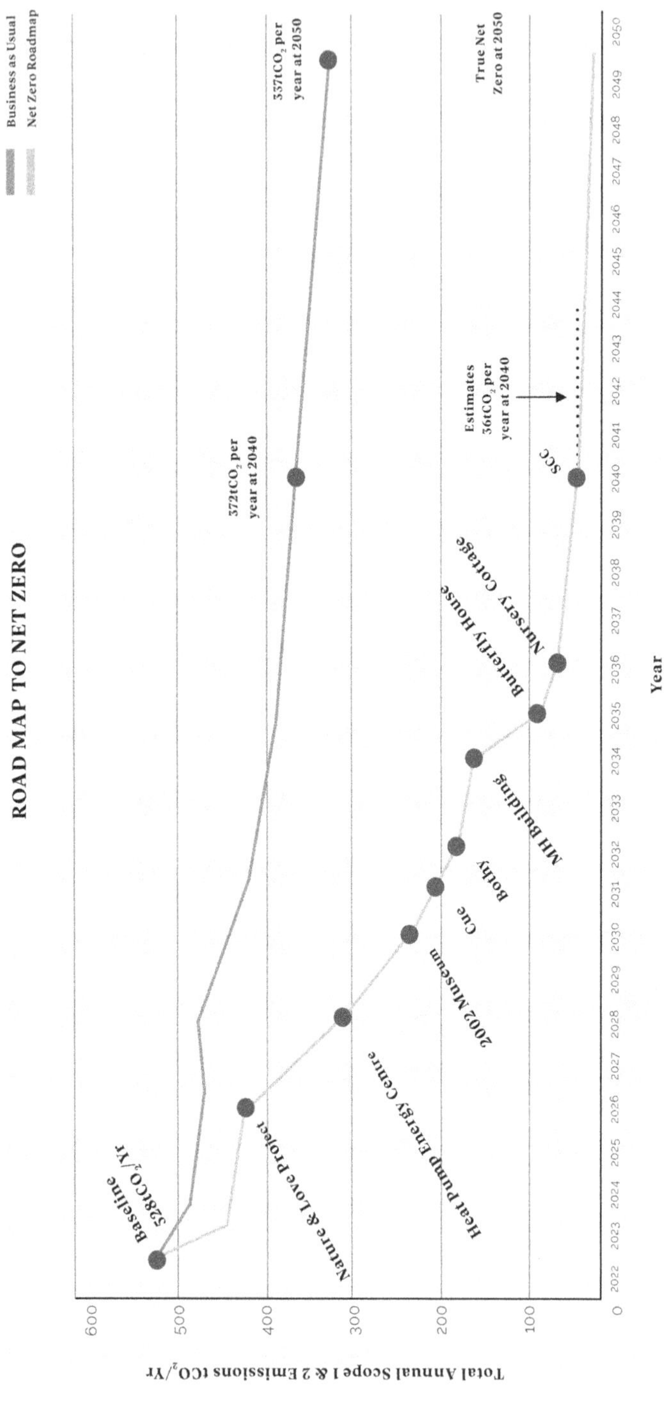

FIGURE 16.1 Horniman road map to net zero. Photo credit: Harley Haddow

greenhouse will be powered by renewable energy sources, and the refurbished Natural History Gallery will reuse or recycle existing infrastructure and the historic showcases. It will create a more stable environment for visitors and collections through more efficient environmental controls, roof insulation, and LED lighting. Solar panels will be installed on the roof, and more efficient boilers, housed in a new energy centre, will aid future transition away from natural gas.

Nature + Love is a ten million pound project funded by the National Lottery Heritage Fund, UK Government and trusts and foundations. Currently, there is no obvious source of funding for all of the other projects. We are in the process of beginning to cost the next few projects in the schedule, so that we can move quickly if and when new sources of funding become available. The UK Government has a legally binding commitment to meet net zero emissions by 2050, so we are anticipating at some point that significant public funding will have to become available to ensure that this target is met.

One of the challenges of reaching GHG neutrality is that it is currently much cheaper to pursue a business as usual model. There is little capital funding or incentive to make the transition to neutrality, and gas remains three times cheaper than electricity. The high inflation levels and extreme energy prices caused principally by the war in Ukraine comprise another disincentive to action. As ever, society – and museums – grapple with short-term demands while trying to progress long-term goals which are in tension with short-term challenges.

Waste reduction

Our waste reduction strategy has consisted of two main areas: to eradicate single-use plastics in our café, shop and supply chain, and both to reduce the volume of waste sent to landfill and to boost the rate of recycling.

By replacement of plastic cups, straws and cutlery with organic-based Vegware and through the eradication of single-use plastic drink bottles, our franchised café has achieved its goal. This has been helped in part by providing a new free drinking fountain near the café to complement the existing Victorian one in the gardens. Our in-house shop has done similarly by replacing plastic items with organic ones, removing all plastic bags, discontinuing all card suppliers with non-compostable wrap and greatly reducing the number of cheaper 'throw away lines' in the pocket money section.

Recycling is a varied story. The bulk of our on-site waste is produced by visitors' consumption of our café products. Although there is a problem with contamination, food waste is mostly recycled through deposition into a Ridan rocket composter which accelerates the composting process. This is added to our garden waste onto a very large compost heap at the foot of the gardens. Overall, 97% of food and garden waste is recycled.

However, despite our café using compostable cups, and extensive signage indicating bins to be used for recycling and for non-recyclable waste, we find it difficult to encourage our visitors to use the correct bins. Food is a major contaminant: food waste in a bin of mixed recycling could potentially lead to the entire contents classed as contaminated and therefore not recycled. Although the café staff sorts most of the customers' waste in the café bins, to avoid contamination, many bins in the rest of the site are contaminated. As a result, our current overall recycling rate in terms of weight is just 27%. We are reviewing ways of addressing this, including the possibility of removing most waste bins from site.

Water

Our Aquarium and our ground-breaking Project Coral make the issue of water use particularly pertinent, and our strategy is to reuse as much water as we can by recycling Aquarium water or harvesting rainwater and using it on our gardens.

The plan is to link the rainwater harvested from the Butterfly House and eventually from the new greenhouse in the Sustainable Gardening Zone (both on the highest part of the site) and store it in large vats. In order for the water to be used for the aquarium, filters will be needed to remove the silt, sediments and leaves coming with the water from the roofs. This cleaner water will then go through the Reverse Osmosis (RO) filtration system which further purifies it for use in the aquarium. Once filtered, the water is piped to the aquarium using gravity. The waste water from the RO will be stored in another tank and be used for toilet flushing and garden watering.

The Nature + Love project will add to this water management strategy in a number of ways. Rain gardens at the upper edge of the natural play zone will increase the contribution to underground aquifers; water attenuation will help prevent flooding by diverting rainwater from going into drains and sewers and will play a part in increasing biodiversity by creating temporary habitats. A rain garden 'furrow' in the centre of the play space will channel water away during rainfall events.

Travel

In order to be an exemplar of an environmentally conscious organisation, the Horniman has tried to encourage both its staff and its visitors to travel in a sustainable manner. In 2020, our staff survey showed that 25% of staff travelled to work by petrol or diesel car or motorbike, and by 2022 this had fallen to 21%. As part of our target to get this to zero by 2040, we have launched a scheme which provides tax incentives to staff to replace their petrol or diesel vehicles with electric ones.

As noted earlier, the issue of whether travel by our visitors to site should be included in our Scope 3 emissions is a vexed one, as it could be argued that if the Horniman didn't exist, visitors would simply go somewhere else rather than save on emissions. However, we wish to do all we can to encourage people to travel sustainably to us, particularly as the commonest method of travel of our visitors remains a private car.

There are a limited number of car park spaces on site reserved for blue badge holders only. The streets around the museum are free to park at the moment, which does not entice visitors away from cars. However, our website explicitly recommends the use of public transport detailing bus, train and bike options. Bike racks for visitors are positioned near the main entrances to the site.

Biodiversity

As noted in the introduction to this book, museums can unify work on climate change and biodiversity which is currently split off into two parallel tracks with their own international conventions and COPs. At the Horniman, the Natural History Gallery and other exhibition spaces are used to engage audiences in positive stories (pp. 249–50), while the gardens and nature trail provide a great opportunity to both increase biodiversity and implement measures that are carbon positive. We have already enjoyed some success with the planting of a Grasslands Garden in 2012 which celebrates critically threatened wild landscapes and introduced to the Horniman plants which are drought-tolerant, which also create stunning wild flower displays that are sustainable and support local wildlife.

We have replaced a number of turfed areas with wildflower meadow planting, and the Bee Garden, opened in 2020, is designed around a central group of six hexagonal raised beds planted with species that attract and provide food for bees. To either side is an area of wildflower turf for maximum plant diversity, with a bee hotel in each – made from reclaimed pallets – to provide shelter and nesting sites for solitary bees.

Throughout the COVID-19 lockdowns, the Horniman Gardens remained open, providing a much-appreciated space for exercise and engagement with nature for local residents, particularly those without gardens of their own. Building on this appreciation, we launched an appeal in 2021 to plant a 'micro-forest' along part of our site that lies adjacent to the very noisy and polluted South Circular Road. It aimed to plant 900 trees in a 300 m^2 space, using the method pioneered by the Japanese botanist Akira Miyawaki. By planting seedlings close together they grow quickly, competing for the light, and the method will rapidly result in a dense micro-forest, able to capture carbon more quickly than traditional reforesting techniques.

We are building on these initiatives with our Nature + Love project. The play area being developed as part of the project is themed around local wildlife (e.g. honeybees, foxes and robins) and will encourage greater care for the environment and biodiversity. Volunteers will help manage the Nature Trail, control invasive species and add more bat and bird boxes as well as monitoring the biodiversity benefits of our interventions. The play area will take inspiration from nature and incorporate natural materials, utilising climbable logs, muddy play, branches for den building, bark paths and repurposing materials where possible. The introduction of a live hedge and hedgerow fence around the play zone will improve biodiversity, and in the long-term carbon sequestration.

The design of the children's café includes a large proportion of solid timber, the lowest embodied carbon option, and will incorporate waterless toilets. It will feature sustainably sourced all-day menus with a vegetarian bias and canned water and plant-based packaging throughout.

Our Sustainable Gardening Zone will include an outdoor engagement space, community garden and interpretation with themes to help community groups and visitors make lifestyle changes, including sustainable habitat creation and choosing sustainably grown and wildlife-friendly plants. Visitors will be able to see the practical application of these as they travel along a step-free route to the new Winter Garden, building on our existing approach to plant for biodiversity and with sustainability and ecology in mind. We will introduce wetland planting where water naturally accumulates at the bottom of the area, improving biodiversity and water attenuation.

Biodiversity on and around the Horniman's site will be increased through biodiversity management plans for the Nature Trail, the Sustainable Gardening Zone and Nature Explorers Adventure Zone and working with partners to create a Green Corridor for Forest Hill.[3] A baseline survey of the whole site in 2022 showed a Defra Metric Biodiversity Calculator score of 13.70 units for area-based habitats and 1.70 units for linear-based habitats (hedgerows). The target is to increase these by 30% for area-based habitats (17.87 units) and by 76% for hedgerows (3.00 units) by the conclusion of the Nature + Love project in 2026.

Training

To meet the challenges of the climate and biodiversity crises, we need a workforce that is familiar with the necessary actions. To help with this, we have delivered carbon literacy training to staff and developed a suite of materials including techniques for improving sustainability in everyday life. This training is mandatory for all staff as well as trainees, apprentices and volunteers involved with the Nature + Love project. It is also available to community groups working on the project – for example those participating in the development of the Sustainable Gardening Zone as well as partner organisations with whom we are working such as Kids in Museums.

The Activity Plan for the Sustainable Gardening Zone in particular focuses on developing horticultural and sustainability skills among our community groups and adult visitors through the creation of a dedicated community garden space as well as an area for workshop delivery.

Engaging the public

As an organisation, the Horniman has the GHG footprint of a medium-sized organisation. It is essential that we do all we can to reduce emissions and waste and to increase biodiversity and act as an example of good practice. However, reaching net zero as an organisation will not make a significant difference in itself to London's overall decarbonisation plans.

With nearly a million visits to site pre-pandemic, and building back to this number as we recover, our main impact will instead be through the extent to which we engage our visitors in making changes to live more sustainably and in adding their voices to the need for change. Our public engagement strategy, therefore, prioritises the climate and biodiversity crises, together with interlinked issues of social justice. Central to this strategy is the need to counter climate crisis anxiety or avoidance by engaging people in positive or hopeful narratives which show that individual actions can aggregate to significant impacts (see Bonanno et al. 2021 for a useful overview). This section gives some examples of how we have done this so far.

Air pollution

The Horniman is located on London's South Circular Road, one of the city's busiest and most polluted roads. Air pollution is one of the most significant issues for the organisation, because of the death in 2013 of a nine-year-old girl, Ella Kissi-Debrah, who lived locally near the road. In 2020, she became the first person in the UK to have air pollution listed as a cause of death by a coroner. As a result, the Mayor of London and the local authority of Lewisham have accelerated measures to curb air pollution. In 2022, when Lewisham was London Borough of Culture, a significant focus of the programme was on this issue, including 'Breathe: 2022' by artist Dryden Goodwin, an ambitious multi-site artwork exploring air pollution combining over 1,000 new drawings, appearing as large-scale still and moving images on sites close to the South Circular Road. As part of this, the Horniman hosted a Day of Community Air Action featuring artist Sarah Stirk's audio-visual installation that seeks to make the invisible threat of pollution visible, hands-on stop motion animation workshops about the call for clean air and the damaging effects of pollution and artists from Climate Museum UK exploring the web of connections between air pollution, other environmental issues and people's experience.

Plastic pollution

With our aquarium and its pioneering coral breeding programme, the issue of plastic pollution in the oceans has loomed particularly large. To engage visitors in a very direct way with this, the aquarium team removed many of the live specimens for two months in 2019 starting on World Oceans Day in an intervention called 'Beat Plastic Pollution'. They were replaced with more than 150 items of single-use and waste plastic, and one of the Horniman's best-loved displays – the jellyfish – was replaced by around 30 plastic bags, commonly mistaken for jellyfish in the ocean and, fatally, eaten by sea turtles. Alongside frogs, fish, seahorses and coral reefs, visitors saw flip-flops hanging from mangrove trees, yogurt pots polluting

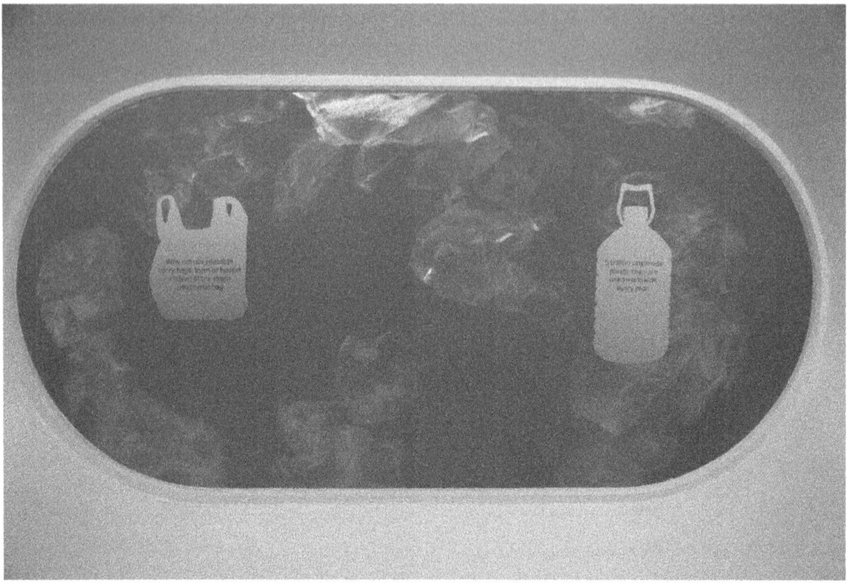

FIGURE 16.2 Jellyfish display replaced by plastic bags in 'Beat Plastic Pollution' intervention. Photo courtesy of Dr Jamie Craggs

the Amazon and drinks bottles lining a British pond. Each exhibit was accompanied by information about the impact plastic has on aquatic creatures and actions visitors could take to reduce this harm.

The intervention was seen by almost 17,500 in-person visitors and many thousands online. Visitors who saw the display were inspired to make environmental pledges, to share and talk about the display, and in the case of one five-year-old, to create a 20-page booklet about plastic pollution which he shared with his school and his MP. In September 2020, the Beat Plastic Pollution display won the Museums and Heritage Award for Limited Budget Project of the Year, which shows the major impact that relatively small interventions and museum programmes can have.

Related to this, we worked a year later with a local school to develop a hanging garden using plastic bottles as a Chelsea Flower Show fringe event. The display featured 120 bottles decorated by the children with designs inspired by French artist Matisse's botanical cut-outs. Each bottle contained plants grown from seed by the children with the help of the Horniman's gardeners, including poppies, marigolds, cornflowers and nasturtiums. The bottles were suspended in groups from the Horniman's library building to form a colourful display.

Exhibitions

Plastic pollution has also featured in our programme of exhibitions in recent years. The 'Inspired By Nature' section of our natural history gallery features interventions by contemporary artists, such as that by Claire Morgan which featured taxidermy specimens of animals found wild on our site (fox, squirrel, blackbird, green parakeet) spewing out plastic debris. Alongside it, the giant turtle display was accompanied by plastic netting and fishing lines. A recent intervention by Harun Morison featured, among other things, the decline of the Jamaican swallowtail butterfly in the face of extensive bauxite extraction on the island (McCormack 2022).

Alongside this, we have featured exhibitions which use art as a method of engaging visitors' emotions rather than as a way of communicating often overwhelming scientific facts. For example 'Meltdown: Visualising Climate Change', a group photographic exhibition curated by Project Pressure, showed the results of commissioning world-renowned artists to conduct expeditions to document changes to the world's vanishing glaciers, the consequences for billions of people and efforts made to limit melting.

It might be thought that climate crisis programming is mainly restricted to adult audiences, but we have felt it vital that we also engage our family visitors. For example 'Monkey Business', a touring family exhibition about primates originated by the National Museum of Scotland, was chosen because alongside sections on behaviour and habitat, it had quite hard-hitting displays on the wildlife trade, habitat destruction and extinction, with a final upbeat section called 'Become an Environmental Hero' which showed children how they could do their bit to help nature and the environment.

Our biggest display project in the next few years is to refurbish the Natural History Gallery (NHG) as the major element in the Nature + Love project. The gallery is some 60 years old, with a high level of interpretation about animal adaptation and classification (much of it now scientifically outdated) and does not mention extinction or climate change. The refurbishment will retain and improve the existing historic cases while changing the content and interpretation and improving the insulation and environmental performance of the gallery.

In terms of content, the new gallery will have sections exploring the colonial and ongoing exploitation of the natural environment and the consequences today in terms of extinction and threatened species. This will be complemented by sections on our own relationship as humans with the natural world of which we are a part, and one on how we need nature, and nature needs us, accompanied by positive stories about action. There will also be a 'Nature Explorers Action Zone' co-created with local families who are not regular museum visitors about engaging people in the environment around us.

Co-curation and engagement

Bringing the voices of the community into our work is increasingly important, as museums move away from an authoritative model of knowledge production to a model of co-produced engagement.

Key to this is the Horniman's Youth Panel, which comprises young people between the ages of 14 and 19 from all over London who meet up regularly to discuss and explore issues central to the museum. Every two years, they mount a display in the World Gallery about an issue that concerns them, with the latest being 'The Sustainable Way' raising awareness of the environmental impact of the cotton industry, pointing out that it takes 2,700 litres of water to make just one cotton T-shirt and that its production is exploitative.

More recently, we have established an Environment Champions community. Initially set up as an online-only 'club' to share information, ideas and tips to empower families with young children to make small steps in their everyday lives to have positive impact on their local environments, it has widened its role and reach to become an umbrella offer for all our environmental events, resources and other activities.

These include targeted community gatherings organised with and without chosen community groups with activities such as litter picking, 'bike doctor' repair sessions, craft activities using natural or recycled materials and outdoor oversize snakes and ladder games on waste and recycling. These onsite activities are complemented by off-site environment-related outreach to engage families from diverse backgrounds in helping to create, manage and preserve more species and habitats on our Nature Trail and in their green spaces and young people from a local estate with gardening projects.

The Facebook group is used to share other community groups' events as well as environmental tips and good news, and a newsletter goes out every other month to 1,366 subscribers.

Conclusion

The Horniman was awarded the Art Fund Museum of the Year prize for 2022, in large part due to its work on the climate and biodiversity crisis. This served as a welcome recognition of the efforts everyone involved in the organisation had been making. Nevertheless, we are only too aware of how much there remains to do, and that we are engaged in a process rather than a finite project. Even if we do reach (almost) GHG neutrality by 2040, we will need to continue efforts to mitigate the impacts of climate change, and to reduce waste and pollution and increase biodiversity. Much of the work remains reliant on external funding, and even measuring progress can be difficult. Our greatest impact will potentially be through motivating our audiences to both make changes in their own lives and demand changes from local and national politicians and businesses. Understanding whether this is happening is difficult, though we will start to get a sense of this as we introduce a new question to our rolling visitor survey, which asks respondents the extent to which they think we are meeting our missions 'to shape a positive future for the world we all share'.

Notes

1 https://aces-org.co.uk/
2 https://www.carbontrust.com/our-work-and-impact/guides-reports-and-tools/conversion-factors-energy-and-carbon-conversion-guide
3 A biodiversity audit of the Nature Trail in 2019 showed 91 species of plant, 19 of birds, three of mammals, three amphibians and a diversity of common invertebrates.

References

Australian Museum. 2019. *Sustainability Action Plan 2019–2021*. Sydney: Australian Museum.

Bonanno, A., Ennes, M., Hoey, J.A., Moberg, E., Nelson, S-M., Pletcher, N. and Tanner, R.L. 2021. Empowering Hope-Based Climate Change Communication Techniques for the Gulf of Maine. *Elementa: Journal of the Anthropocene*, 9(1): 13–28. https://doi.org/10.1525/elementa.2020.00051

Hodgson, C. and Nauman, B. 2021. Carbon Offsets: A License to Pollute or a Path to Net Zero Emissions? *Financial Times*, 31 August.

Julie's Bicycle. 2019. *Tate Carbon Footprint*. London: Julie's Bicycle.

Julie's Bicycle. 2021. *The Art of Zero*. London: Julie's Bicycle.

McCormack, C. 2022. Staying Afloat. Harun Morison Interviewed by Chris McCormack. *Art Monthly*, 1 September 2022.

V&A. 2021. *Sustainability Plan 2021–2024*. London: Victoria and Albert Museum.

WTTC. 2021. *A Net Zero Roadmap for Travel & Tourism: Proposing a New Target Framework for the Travel & Tourism Sector*. New York: World Travel and Tourism Council.

INDEX

For Product Safety Concerns and Information please contact our EU
representative GPSR@taylorandfrancis.com
Taylor & Francis Verlag GmbH, Kaufingerstraße 24, 80331 München, Germany

www.ingramcontent.com/pod-product-compliance
Ingram Content Group UK Ltd.
Pitfield, Milton Keynes, MK11 3LW, UK
UKHW020149200926
475702UK00016B/2619
* 9 7 8 1 0 3 2 3 8 9 4 1 7 *